Friendly Chemistry

A Guide to Learning Basic Chemistry
4th Edition

By Joey and Lisa Hajda

Student Edition

Published by Hideaway Ventures, 79372 Road 443, Broken Bow, NE 68822

For additional information regarding this publication please contact Joey or Lisa Hajda at the above address or call 308-870-4686, email hideaway1@gpcom.net or visit our website: www.friendlychemistry.com

Copyright 1996, 1998, 2008, 2010 Hideaway Ventures, Joey or Lisa Hajda

All rights reserved. No portion of this book may be reproduced in any form, or by any electronic, mechanical or other means, without the prior written permission of the publisher.

Friendly Chemistry

Table of Contents

	Page
Lesson 1: Meet Chemistry	S1
Lesson 2: Moving in a Little Closer	S19
Lesson 3: The Arrangement of Electrons in Atoms	S27
Lesson 4: Applying Quantum Mechanics	S37
Lesson 5: Learning the Code: Orbital Notation	S45
Lesson 6: Electronic Configuration Notation	S53
Lesson 7: Electron Dot Notation	S61
Lesson 8: Relating Electron Arrangement to Reactivity	S69
Lesson 9: Another Trend of the Periodic Table: Ionization Energy	S81
Lesson 10: A Final Trend of the Periodic Table: Atom Size	S89
Lesson 11: Ion Formation	S97
Lesson 12: Determining Charges on Ions	S107
Lesson 13: Forming Compounds from Ions	S117
Lesson 14: Learning Slightly More Complex Ions	S125
Lesson 15: Making Compounds with Polyatomic Ions	S135
Lesson 16: Introducing the Mole	S139
Lesson 17: Finding Formula Weights	S147
Lesson 18: Finding the Percent Composition of Compounds	S157
Lesson 19: Writing Empirical Formulas	S165
Lesson 20: Putting Compounds into Reactions	S183
Lesson 21: Balancing Chemical Reactions	S201

Lesson 22: Introduction to Stoichiometry……………………...……………...S213

Lesson 23: Predicting Grams of Product……………….…………………..S223

Lesson 24: Predicting Grams of Product from Grams of Reactant…………….S231

Lesson 25: Do I Have Enough Reactant?..S239

Lesson 26: Mixing Compounds with Water to Make Solutions……………….S247

Lesson 27: Incorporating Molarity into Stoichiometric Problems………………S255

Lesson 28: Determining the Needed Amount of a Solution to Perform a Reaction…S265

Lesson 29: Properties of Gases and How Gases can be Measured in Reactions……S275

Lesson 30: How Changing Temperatures Affect Gases…………………….S285

Lesson 31: How Changing Pressures Affect Gases…………………………S295

Lesson 32: The Combined Gas Laws………………………………..……...S301

Index and Useful Conversions……………………………………..……...S309

Periodic Table of the Elements……………………………………….……S311

Lesson 1: Meet Chemistry!

When you meet someone for the first time, you like to know that person's name and something about him or her. Since we are getting acquainted with chemistry, let's get to know the subject a little better by learning where the word *chemistry* originated. In about 100 AD, Greek scientists were very busy studying scientific processes in an attempt to change non-valuable elements into more valuable elements such as gold. The Greeks thought elements naturally "transmutated" into gold in the earth, and these scientists wanted to learn those "transmutation" processes and be able to repeat them in the lab. These theories of turning simpler, more common elements into gold (known as *alchemy*), were also taking place in China and other locations. The popularity of the idea rose and declined over several hundred years and although it eventually was found to be impossible, many, many ideas and processes were discovered about the nature of the earth's elements. It is from the term *alchemy* that our present-day term of *chemistry* is derived.

Now that you know where your new friend's "name" came from, let's get to know more about chemistry. It is generally taught that chemistry is the study of matter (pretty simple, so far) and the way various kinds of matter react with each other (still fairly simple!). Matter is defined as any substance, whether it be solid, liquid or gas. And that is basically what chemistry is all about!

Now, you might say, "Sure, that definition sounds so simple and easy to understand, but what about all those neutrons and isotopes and all those symbols and foreign-looking codes for matter?" We have to admit that there is almost another language you will begin to learn as you study chemistry. With this book, you will learn a great deal about matter and the way different types of matter react with each other *as well as* the words and symbols used to describe those substances and their reactions. As you continue through this book, you will be introduced to new terms and symbols that will soon become second nature. Expect to find yourself using more and more of the terms we discuss! So, for now, let's just stick with our simple definition of **chemistry: the study of matter and how various kinds of matter react with each other.**

When we speak of matter, especially in the context of learning chemistry, visions of bottles containing strange-smelling crystals and colored liquids may come to mind. However, the matter we are referring to is everywhere around you! Your notebook, your pencil, you, your room, your house, your food, and the very vital substance required by all living things – water – are kinds of matter that we can study in the context of chemistry. Chemistry is not reserved only for the study of those odd-smelling crystals and liquids. Chemistry can be applied to any object around you. The wood in your house can be analyzed and found to be composed of carbon, hydrogen and oxygen. The hamburger you enjoy is also made of carbon, hydrogen and oxygen with some added nitrogen. And, as we have already said, the water that you drink and wash with (which is made of hydrogen and oxygen) is one of the most "down to earth" kinds of matter that we could discuss. So, if your vision of chemistry was bubbling liquids in corkscrew-shaped tubes being monitored by people wearing goggles and white coats, alter it slightly to include almost everything around you!

Did you catch some "scientific language" in the preceding paragraph (hydrogen, oxygen, carbon and nitrogen)? Those are names of **elements**.

Ancient chemists began to understand that there were certain kinds of existing matter, that could not be broken down into simpler forms of matter. These forms of matter, that could not be separated by one means or another, were given the name **elements**, indicating that they were elemental or elementary or the basis for all that follows, as in elementary school. Combined elements are what make up matter. Examples of elements that you are probably familiar with are hydrogen, oxygen, lead and gold. There are more than 100 known elements today. The actual number is difficult to say since new elements are being discovered or synthesized as you read this. Look at table on the next page to see a list of currently known elements. Note that 92 of those elements are considered to be naturally occurring elements; that is, to occur on earth, not having been made by man. Note that the rest are considered not to be naturally occurring since these elements have been made by scientists.

The history of naming elements is very interesting and a study unto itself! Some names and symbols may appear to be strange and obscure. You will find in the examples we use to illustrate concepts that many of the same elements are mentioned over and over again. You will pick up names and symbols of the more common elements as we go along.

Look now at the periodic table of elements found on the next page. Look first at the key which shows the information found in each square of the table. Note how the element symbol consists of 1-3 letters and that the first letter is always an upper case letter. If there is more than one letter for a symbol, the second and third letters are always lower case letters. Observe the numbers found in each square. Note how the atomic number is always a whole number and the atomic mass number is *not* a whole number. We will learn much more about these number later in the course.

Let's review what has been discussed so far. We first stated that **chemistry** is the study of matter and how various kinds of matter react with each other.

Second, everything around us is composed of matter and the study of matter could be applied to all things.

Finally, we learned that some matter cannot be broken down into simpler forms of matter and is designated as an **element**.

S3

Friendly Chemistry

Element Name	Symbol	Element Name	Symbol	Element Name	Symbol
Actinium*	Ac	Hafnium*	Hf	Proactinium*	Pa
Aluminum*	Al	Hahnium	Ha	Radium*	Ra
Americum	Am	Helium*	He	Radon*	Rn
Antimony*	Sb	Holmium*	Ho	Rhenium*	Re
Argon*	Ar	Hydrogen*	H	Rhodium*	Rh
Arsenic*	As	Indium*	In	Rubidium*	Rb
Astatine*	At	Iodine*	I	Ruthenium*	Ru
Barium*	Ba	Iridium*	Ir	Rutherfordium	Rf
Berkelium	Bk	Iron*	Fe	Samarium*	Sm
Beryllium*	Be	Krypton*	Kr	Scandium*	Sc
Bismuth*	Bi	Lanthanum*	La	Selenium*	Se
Boron*	B	Lawrencium	Lr	Silicon*	Si
Bromine*	Br	Lead*	Pb	Silver*	Ag
Cadmium*	Cd	Lithium*	Li	Sodium*	Na
Calcium*	Ca	Lutetium*	Lu	Strontium*	Sr
Californium	Cf	Magnesium*	Mg	Sulfur*	S
Carbon*	C	Manganese*	Mn	Tantalum*	Ta
Cerium*	Ce	Mendelevium*	Md	Technetium*	Tc
Cesium*	Cs	Mercury*	Hg	Tellurium*	Te
Chlorine*	Cl	Molybdenum*	Mo	Terbium*	Tb
Chromium*	Cr	Neodymium*	Nd	Thallium*	Tl
Cobalt*	Co	Neon*	Ne	Thorium*	Th
Copper*	Cu	Nickel*	Ni	Thulium*	Tm
Curium	Cm	Nitrogen*	N	Tin*	Sn
Dysprosium	Dy	Nobelium	No	Titanium*	Ti
Einsteinium	Es	Osmium*	Os	Tungsten*	W
Erbium*	Er	Oxygen*	O	Uranium*	U
Europium*	Eu	Palladium*	Pd	Vanadium*	V
Fermium	Fm	Phosphorus*	P	Xenon*	Xe
Fluorine*	F	Platium*	Pt	Ytterbium*	Yb
Francium*	Fr	Plutonium*	Pu	Yttrium*	Y
Gadolinium*	Gd	Polonium*	Po	Zinc*	Zn
Gallium*	Ga	Potassium*	K	Zirconium*	Zr
Germanium*	Ge	Praeseodymium*	Pr		
Gold*	Au	Promethium*	Pm		

* indicates naturally occurring element

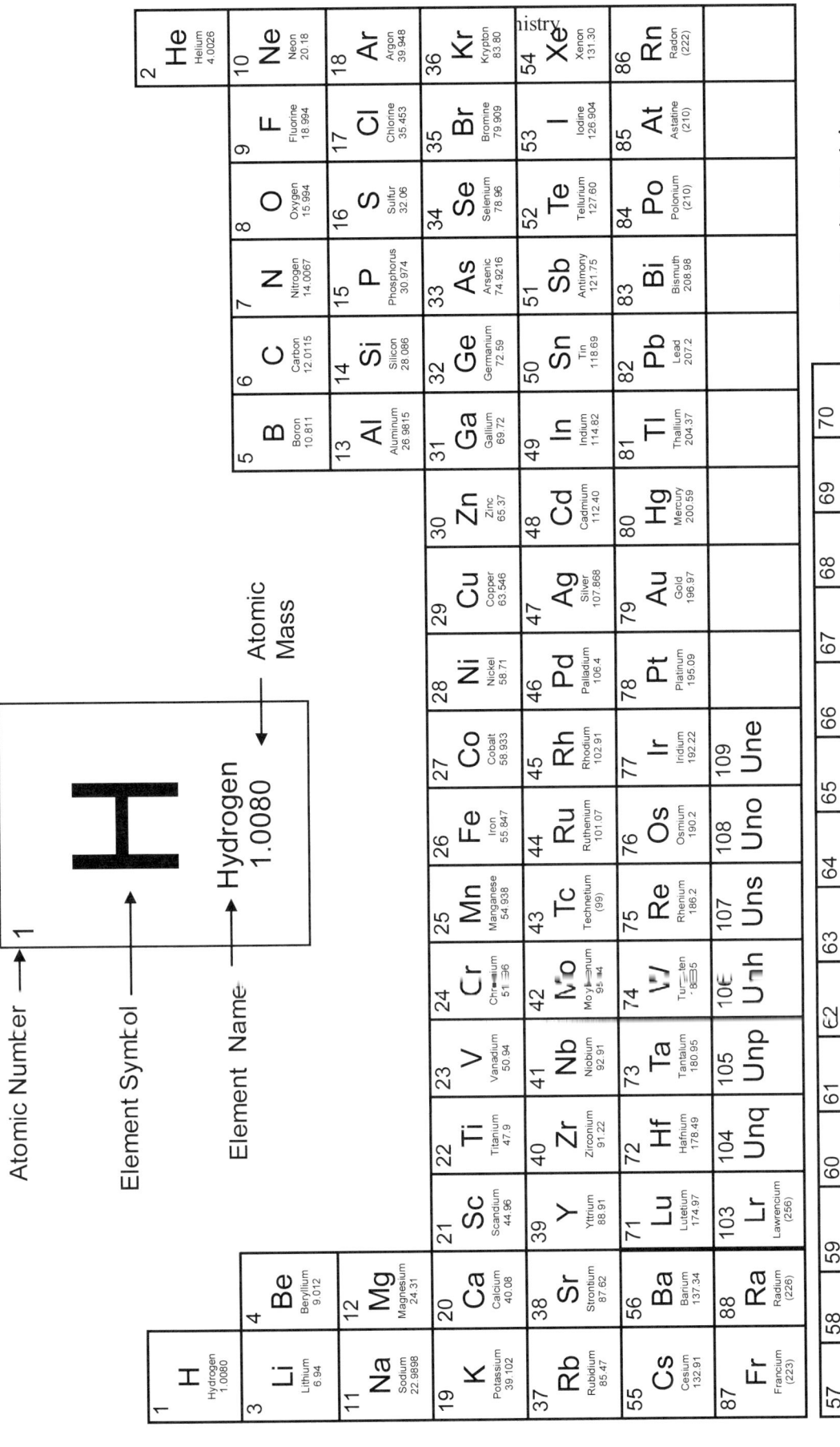

Periodic Table Of the Elements

Friendly Chemistry

Element Bingo

Below is a blank bingo card. In each square, write the symbol of an element. Write that symbol only once. When everyone has done this, your teacher will begin calling out the names of elements. When an element is called, look for its symbol on your card. Place a marker (colored chip, button, etc.) on that square. When you get six in a row across, down or diagonally, yell "BINGO," to potentially win the round. In order to win, you must correctly identify the elements whose symbols you marked.

He					

Friendly Chemistry

Element Bingo 2

Play two cards at once! Fill in this card like you did the first. Double your chances to win that nifty prize your teacher has for the winner.

Ne					

Friendly Chemistry

Name_____ Date_____

Friendly Chemistry

Element Symbol Practice 1

Write the correct element symbol beside each element name below. Some questions may appear more than once. The first problem has been completed for you!

1. Lithium	Li	21. Manganese	
2. Magnesium		22. Magnesium	
3. Carbon		23. Chlorine	
4. Hydrogen		24. Silicon	
5. Oxygen		25. Phosphorus	
6. Helium		26. Iron	
7. Boron		27. Cobalt	
8. Fluorine		28. Copper	
9. Calcium		29. Nickel	
10. Nitrogen		30. Arsenic	
11. Strontium		31. Argon	
12. Sodium		32. Aluminum	
13. Neon		33. Gallium	
14. Copper		34. Germanium	
15. Beryllium		35. Rubidium	
16. Argon		36. Silver	
17. Sulfur		37. Beryllium	
18. Chromium		38. Selenium	
19. Oxygen		39. Bromine	
20. Potassium		40. Iodine	

Friendly Chemistry

Name_____ Date_____

Friendly Chemistry

Element Symbol Practice 2

Write the correct element name beside each element symbol below. Some questions may appear more than once. The first problem has been completed for you!

1. C	Carbon	21. Mg		41. K	
2. Al		22. N		42. He	
3. He		23. K		43. N	
4. H		24. Si		44. F	
5. Be		25. Zn		45. Na	
6. Fe		26. Zr		46. Ne	
7. F		27. Al		47. Ni	
8. Cl		28. Cr		48. N	
9. Ca		29. As		49. C	
10. Ni		30. Ar		50. Ca	
11. Br		31. B		51. Cl	
12. Si		32. Ga		52. Co	
13. Ne		33. Ge		53. Cu	
14. Cr		34. Rb		54. Cr	
15. V		35. O		55. Ar	
16. B		36. Ag		56. As	
17. Na		37. Se		57. Mg	
18. Cu		38. Sn		58. Mn	
19. S		39. Br		59. S	
20. I		40. H		60. Au	

Name_____ Date_____

Friendly Chemistry

Element Symbol Practice 3

Circle the question number for the element name and symbols which are written correctly. The first question has been completed for you!

1.	(C)	Carbon	21.	Mg	Magnesium
2.	Al	Alluminum	22.	N	Nitorgen
3.	He	Heleum	23.	K	Potasium
4.	H	Hydrogen	24.	Si	Silicon
5.	Be	Beryllium	25.	Zn	Zirconium
6.	Fe	Iron	26.	Zr	Zinc
7.	F	Flourine	27.	Li	Lithium
8.	Cl	Clhorine	28.	Cr	Chromium
9.	Ca	Carbon	29.	As	Arsenic
10.	Ni	Nickel	30.	Ar	Argon
11.	Br	Bromine	31.	B	Boron
12.	Si	Silver	32.	Ga	Galium
13.	Ne	Neon	33.	Ge	Germium
14.	Cr	Chromium	34.	Rb	Rubidium
15.	Va	Vanadium	35.	O	Oxygin
16.	Be	Berkelium	36.	Ag	Silver
17.	Na	Sodium	37.	Se	Selenium
18.	Cu	Cobalt	38.	Sn	Tin
19.	S	Sodium	39.	Br	Barium
20.	I	Iron	40.	Hf	Hafnium

S10

Friendly Chemistry

Lesson 1 Lab Investigation

In order to make your first lab fun and meaningful (and safe!), please read the following first:

What you're going to do:

In your first lab activity, you will be given several "unknown powders" which you will be asked to identify. By making observations of these powders using your five senses, hopefully you'll be able to determine the identity of each!

To help keep things organized, you have been supplied with data tables on which to record your observations for each unknown powder. You will find these on the next pages. Note that each unknown powder gets its own data table.

Turn the page and look at one of those data tables now. Note that there is a place to put the ID number for each powder. Also note that you will systematically make the same observations for each powder (how each looks, smells, feels, sounds and tastes).

After these initial observations you will test a portion of each by adding vinegar and then some iodine solution. DO NOT taste any of these powders once you have added the vinegar or iodine solution!!! Vinegar is an acid and is very sour and iodine can be poisonous!!! DO NOT taste any powder after adding the vinegar or iodine solution!

Your instructor may ask that you wear appropriate safety gear as you conduct this lab activity. We strongly suggest you wear safety glasses (not only is vinegar sour, but it can really sting if you get it in your eyes!). Note also that iodine can stain your skin and clothes. Wearing an apron or smock would be a good idea, too!

When you're done:

Follow your instructor's instructions to clean up your lab area. Then be ready to discuss what you think each powder could be. Check with your instructor if he or she would like a written report.

Remember:

Be careful;
Do only what your instructor will allow, but ask if you'd like to try something;
Help get things cleaned up; and
Have fun!

Friendly Chemistry

Data Table for Unknown Powders Lab

Name_____

Unknown Powder ID Number _____

	When I first looked at this unknown powder...	When I added drops of water...	When I added drops of vinegar ...	When I added drops of iodine ...	Additional observations I made...
it **LOOKED** like...					
it **SMELLED** like...					
it **FELT** like...					
it **SOUNDED** like...					
it **TASTED** like...		DO NOT TASTE AFTER ADDING TEST SOLUTIONS!			

Based upon my observations, I hypothesize that this substance is _____.

S12

Friendly Chemistry

Data Table for Unknown Powders Lab

Name_____

Unknown Powder ID Number _____

	When I first looked at this unknown powder...	When I added drops of water...	When I added drops of vinegar ...	When I added drops of iodine ...	Additional observations I made...
it LOOKED like...					
it SMELLED like...					
it FELT like...					
It SOUNDED like...					
it TASTED like...		DO NOT TASTE AFTER ADDING TEST SOLUTIONS!				

Based upon my observations, I hypothesize that this substance is _____.

Friendly Chemistry

Data Table for Unknown Powders Lab

Name _____

Unknown Powder ID Number _____

	When I first looked at this unknown powder...	When I added drops of water...	When I added drops of vinegar ...	When I added drops of iodine ...	Additional observations I made...
it LOOKED like...					
it SMELLED like...					
it FELT like...					
it SOUNDED like...					
it TASTED like...		DO NOT TASTE AFTER ADDING TEST SOLUTIONS!			

Based upon my observations, I hypothesize that this substance is _____.

S14

Friendly Chemistry

Data Table for Unknown Powders Lab

Name_____

Unknown Powder ID Number _____

	When I first looked at this unknown powder...	When I added drops of water...	When I added drops of vinegar ...	When I added drops of iodine ...	Additional observations I made...
it LOOKED like...					
it SMELLED like...					
it FELT like...					
It SOUNDED like...					
it TASTED like...		DO NOT TASTE AFTER ADDING TEST SOLUTIONS!			

Based upon my observations, I hypothesize that this substance is _____.

Friendly Chemistry

Data Table for Unknown Powders Lab

Name _____

Unknown Powder ID Number _____

	When I first looked at this unknown powder...	When I added drops of water...	When I added drops of vinegar ...	When I added drops of iodine ...	Additional observations I made...
it **LOOKED** like...					
it **SMELLED** like...					
it **FELT** like...					
it **SOUNDED** like...					
it **TASTED** like...		DO NOT TASTE AFTER ADDING TEST SOLUTIONS!			

Based upon my observations, I hypothesize that this substance is _____.

Friendly Chemistry

Data Table for Unknown Powders Lab

Name_____

Unknown Powder ID Number _____

	When I first looked at this unknown powder...	When I added drops of water...	When I added drops of vinegar ...	When I added drops of iodine ...	Additional observations I made...
it **LOOKED** like...					
it **SMELLED** like...					
it **FELT** like...					
it **SOUNDED** like...					
it **TASTED** like...		DO NOT TASTE AFTER ADDING TEST SOLUTIONS!			

Based upon my observations, I hypothesize that this substance is _____.

S17

Friendly Chemistry

Data Table for Unknown Powders Lab

Name _____

Unknown Powder ID Number _____

	When I first looked at this unknown powder...	When I added drops of water...	When I added drops of vinegar ...	When I added drops of iodine ...	Additional observations I made...
it LOOKED like...					
it SMELLED like...					
it FELT like...					
it SOUNDED like...					
it TASTED like...		DO NOT TASTE AFTER ADDING TEST SOLUTIONS!			

Based upon my observations, I hypothesize that this substance is _____.

Lesson 2: Moving in a Little Closer

In Lesson 1, we said that chemistry is the study of matter and that matter is all around us. Find something near you right now that is made of matter. Write down in the following space what you chose.

The piece of matter I chose is _____.

Now, in the space below, write down to the best of your knowledge what that piece of matter is made up of or, in other words, the ingredients that go together to make it what it is. Here is an example: Let's say I chose *table* as my piece of matter. I could write down that the *table* is made of *wood, glue, screws, paint and varnish, some bits of felt under the legs and hinges where the leaves drop down.* Get the idea?

Ingredients which make up my piece of matter: _____

_____

Now, choose one component from your list of ingredients. For our example of the table, I am going to choose *glue*. Write down your choice here.

My choice is _____.

Carefully think about that one ingredient and then write down what you think makes up *that* ingredient. In the example of *glue,* I could write *bones, cow hooves and water*. (Please don't worry if you get stumped at this point in this exercise. Just follow along with the example!).

The ingredients which make up this ingredient are: _____
_____

Now, choose one of those ingredients that you just listed. (You're probably saying to yourself, "Will this ever end?") For our example I am going to choose the *water*.

Now, you guessed it, write down what ingredients make up that ingredient! I would write down that water is made up of hydrogen and oxygen, and you might know that, too! Can you write down the ingredients of the ingredient of the ingredient you chose?

The ingredients which make up this ingredient are: _____
_____

As you can obviously see, we are getting into smaller and smaller bits of matter and, in the case of our example, we got down to saying that the water was made up of hydrogen and oxygen. Do the names hydrogen and oxygen ring any bells? Yes, those are the names of two elements we used in the last chapter!

In our example, we eventually got to the point where we found ourselves naming elements (hydrogen and oxygen) as the ingredients of our ingredients of our ingredients. According to our definition, an element is matter that cannot be broken down into a simpler form of matter; that it is a most basic form from which other things were made. However, in our study of chemistry, we cannot stop here! In order to understand more fully the nature of matter and how that matter reacts with other matter (have you heard this definition somewhere before?) we must learn about what constitutes an element. In other words, what are the ingredients, or components, of an element?

You may know that the "ingredients" of an element are **atoms**. Atoms are what make up elements. The early Greek scientists who devised this theory (yes, the information upon which we will base the rest of our study of basic chemistry *is* theory) used the term *atoms* meaning "in-divisible" or unable to divide again. **This theory that all matter is made up of elements which are made of atoms is known as the Atomic Theory.**

Let's pause and review for a moment. In this lesson, we began with an exercise in which we took an object made of matter and proceeded to analyze its ingredients. We then found the ingredients of the ingredients and on and on until we got to the point where the list of ingredients was composed of elements. We had learned earlier that elements were the simplest forms of matter. We also said that to fully understand chemistry (the study of matter and how that matter reacts with other matter), we had to go one more step. That step was to study the atoms that make up those elements and that this study would be based upon the Atomic Theory. Let's move on now!

In the last lesson we discussed how all matter is made of elements which are made of atoms. To fully understand matter and how it reacts with other matter (definition of chemistry, right?), we need to learn more about the "ingredients" of even atoms. This lesson is designed to introduce you to the three basic ingredients of any atom: **neutrons, protons and electrons**. We can give these three components a scientific name: **subatomic particles**. We will not go into great depth about the origins of the theories of subatomic particles. Like the development of the Atomic Theory, the developments of these theories are interesting. If you would like to learn more, please consult the chemistry section of your local library.

Let's begin our study of atoms by considering the general appearance of an atom. It is believed that atoms are like mini-solar systems. The **nucleus** in the center of the atom corresponds to the sun in the center of our solar system. The **electron cloud** corresponds to the planets circling the sun. Within the nucleus (the "sun") are found two of the subatomic particles: the protons and the neutrons. Circling about the nucleus is an arrangement of electrons (the "planets," in our mini-solar system analogy). The electrons are considered to be moving at extremely high speeds. Therefore, if you could see one atom, all you would see of its electrons is a blur, hence the name electron cloud. Look at the diagram of an atom on the next page.

Friendly Chemistry

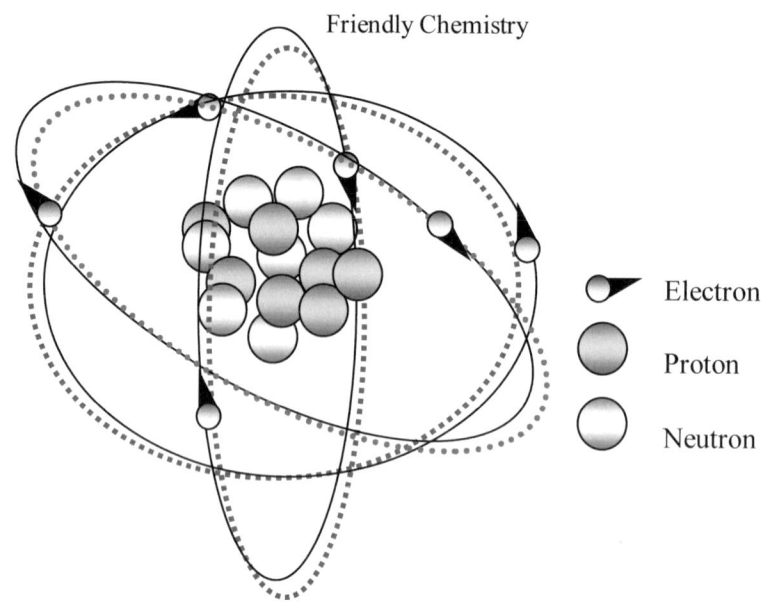

We said that within the nucleus you would find two of the subatomic particles: the protons and the neutrons. Although the neutrons and protons are extremely tiny particles, they make up most of the amount of mass of an atom. **Mass is the amount of matter in an object.** The electrons, on the other hand, make up only a small fraction of the mass of an atom. Let's look at some characteristics of these subatomic particles to get an idea how small they are thought to be and their approximate mass.

Subatomic Particle	Actual Mass	Relative Mass
Protons	1.7×10^{-24} grams	Cannon ball
Neutrons	$\sim 1.7 \times 10^{-24}$ grams	Cannon ball
Electrons	9.1×10^{-28} grams	Ping-pong ball

Note the $10^{-28\text{th}}$!!

Note from the values in the table above that almost all of the mass of an atom is found in its protons and neutrons! The electrons have an extremely small amount of mass!

The number of subatomic particles in any one atom is the basis for the **atomic number** of the atom. Look at the periodic table on the next page 2-16. Note that in each little square you can find an element symbol, an element name, and two numbers. The number in the upper left-hand corner (the whole number) is the atomic number. The number beneath the element symbol and name is the **atomic mass** of the element.

Find the symbol 'C' for carbon. Carbon's atomic number is 6 which means that one atom of carbon contains six protons, six neutrons and six electrons   This holds true for the rest of the elements in the periodic table. If you could sample a few atoms from any element in the periodic table, all of the atoms in that sample would generally have the same number of protons, neutrons and electrons. We say "generally" because the number of neutrons can vary slightly. All in all, we can say that the atomic number is (most of the time) equal to the number of protons or electrons or neutrons for that atom. Try some of these exercises:

1. What is the atomic number for helium? _____
2. How many protons does an atom of helium have? _____
3. How many neutrons does an atom of helium generally have? _____
4. How many electrons does an atom of helium have? _____

[If you answered '2' for each question above, you're correct! If not, be sure you found the correct element in the periodic table. The symbol for helium is 'He' and can be found in the upper right-hand corner of the periodic table.]

5. How many protons does oxygen (O) have? _____
6. How many neutrons does oxygen generally have? _____
7. How many electrons does oxygen have? _____
8. How many protons does sulfur (S) have? _____
9. How many protons does magnesium (Mg) have? _____
10. How many electrons does argon (Ar) have? _____

Let's review now what we've learned in this lesson:
- Tiny bits of matter are called atoms.
- Atoms are thought to look like tiny solar systems with a nucleus in the center with an electron cloud around it.
- Atoms are made up of subatomic particles: protons, neutrons and electrons.
- Protons and neutrons are found in the nucleus of an atom. Electrons are found in orbits around the nucleus of the atom.
- The atomic number of an atom tells how many protons, neutrons and electrons are present in that element.
- Mass is the measure of matter in an object.
- Most of the mass of an atom is found in its protons and neutrons. Electrons have a very small mass.

Periodic Table Of the Elements

Friendly Chemistry

Friendly Chemistry

Name_____ Date_____

Friendly Chemistry

Element Symbol Practice

Write the correct element symbol beside each element name below. Some questions may appear more than once. The first problem has been completed for you!

1. Lithium	Li	21. Molybdenum	
2. Manganese		22. Copper	
3. Calcium		23. Chlorine	
4. Hydrogen		24. Silicon	
5. Argon		25. Phosphorus	
6. Arsenic		26. Iron	
7. Boron		27. Cobalt	
8. Fluorine		28. Helium	
9. Cobalt		29. Nickel	
10. Nitrogen		30. Arsenic	
11. Nickel		31. Argon	
12. Sodium		32. Aluminum	
13. Neon		33. Gallium	
14. Bromine		34. Germanium	
15. Beryllium		35. Rubidium	
16. Arsenic		36. Silver	
17. Sulfur		37. Beryllium	
18. Chromium		38. Selenium	
19. Zinc		39. Bromine	
20. Potassium		40. Tin	

Friendly Chemistry

Name_____ Date_____

Friendly Chemistry

Subatomic Particle Numbers Practice

Below is a table which lists information about elements on the periodic table. Use the provided clues to fill in the missing information.

Element Symbol	Element Name	Number of Protons	Number of Neutrons	Number of Electrons
1. C				
2.	Sulfur			
3.		8		
4.			12	
5.				3
6. Mn				
7.	Neon			
8.		11		
9.			13	
10.				15
11. Cl				
12.	Argon			
13.		19		
14.			30	
15.				2
16. Br				
17.	Iodine			
18.		21		
19.			1	
20.				5

S26

Lesson 3: The Arrangement of Electrons in Atoms

In Lesson 2 we discussed the names and locations of the subatomic particles found in atoms. We stated that the atomic number for any element found on the periodic table tells the number of protons or electrons and "generally" the number of neutrons found in each atom of that element. Let's continue now and learn how the electrons are arranged in atoms.

First, let's examine the protons and electrons a bit more closely. Based upon experiments done by some famous chemists, it is accepted that electrons and protons carry an electrostatic charge. You might think of an electrostatic charge as being like the tiny spark you can elicit after you shuffle your feet across a carpet and then touch a doorknob. In this example, the spark you see (and can even feel) is negatively charged electrons moving from you through the air to the doorknob. On the atomic level, **electrons carry a negative electrostatic charge**.

On the other hand, **protons carry a positive electrostatic charge**. Neutrons carry a neutral (or no) charge. With this in mind, we can see that the nucleus, since it is composed of protons and neutrons, in effect, carries an overall **positive** charge while the electron cloud carries a **negative** charge. It is theorized that these opposite charges (positive nucleus and negative electron cloud) attract each other like the north and south poles of magnets attract each other. We can make an analogy in which the electrons are held in an orbit around the nucleus similar to the way the planets (or any other satellites) are held in orbit around the sun due to gravitational forces. See the diagrams below. The electrons, like the planets, have energy of motion that keeps them moving in an orbit around the nucleus. If you would like to learn more details about the experiments mentioned above, please consult the chemistry section of your local library.

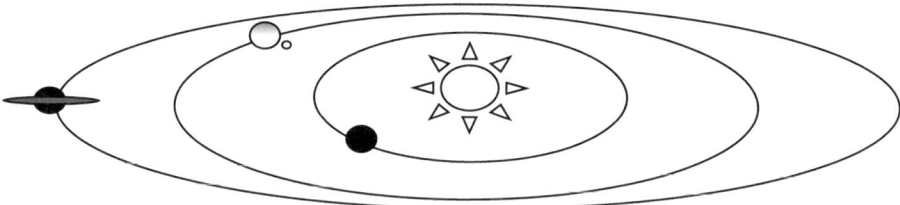

Planets are held in orbit by gravitational forces from the sun.

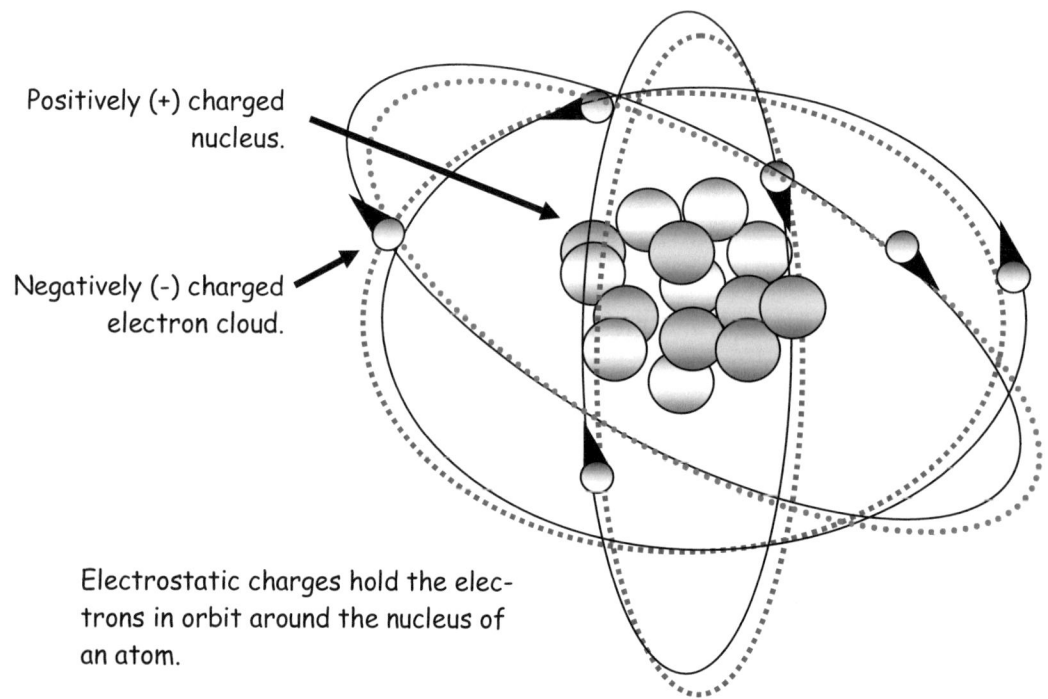

Positively (+) charged nucleus.

Negatively (-) charged electron cloud.

Electrostatic charges hold the electrons in orbit around the nucleus of an atom.

While the neutrons and protons are clustered together in the nucleus, the electrons are thought to be in various arrangements circling the nucleus. It is this variation in arrangement of the electrons that is responsible for the way each element interacts with other elements. In other words**, the arrangement of electrons around each nucleus of an element's atoms determines the reactivity of that element.** As you might suspect, some elements are highly reactive while some appear to not react at all with others. Consider the fate of the Hindenbergh. The hydrogen-filled derigible of the early 1900's crashed, which resulted in the hydrogen reacting with oxygen. The resulting raging fire consumed the entire airship. Compare the Hindenbergh's hydrogen with helium which is used today in blimps and party balloons. Helium is very non-reactive compared to hydrogen or even pure oxygen, for that matter!

Sodium, as another example, is so reactive with water that is must be stored within a solution of diesel to keep the explosive metal from coming into contact with water vapor droplets in the air! The arrangement of those tiny electrons makes an enormous (and sometimes explosive) difference in how elements react with other elements.

Let's move on and learn more about how these electrons are arranged and how the arrangements affect an atom's reactivity. There is an area of scientific knowledge used to describe the arrangement of electrons about an atom's nucleus. This is known as **quantum mechanics**. Quantum mechanics is just a means for describing where the electrons are, kind of like a code which can help you understand more about the atoms that make up elements.

The code of quantum mechanics is based upon four quantum numbers (although, only the first quantum number is actually a number while the other three are letters or symbols). You might think of these four quantum numbers as being like the letters in our alphabet: a set of symbols which, together, make up words which have meaning. The four quantum numbers, when used together, create a meaningful description of the arrangement of an atom's electrons.

The first quantum number is known as the **principle quantum number**. As its name implies, the principle quantum number serves as a basis upon which the other three quantum numbers are built. The principle quantum number, as we stated above, is actually a number, and the principle quantum number is a whole number. The lowest principle quantum number is 1 and the highest is 7. Each principle quantum number corresponds to the position the electrons occupy as they travel around the nucleus of the atom. In other words, not all of the electrons in an atom travel in the same path at the same distance from the nucleus. They are arranged in concentric orbits or energy levels around the nucleus just like the planets in our solar system revolve around the sun at various distances. The first orbit (or energy level) from the nucleus is given the principle quantum number of 1. The second orbit (or energy level) moving outward from the nucleus is given the principle quantum number of 2. The third orbit or (energy level) is given the principle quantum number of 3 and so on up until you reach the seventh level which is given-yes, you've got it-the principle quantum number of 7. You might think

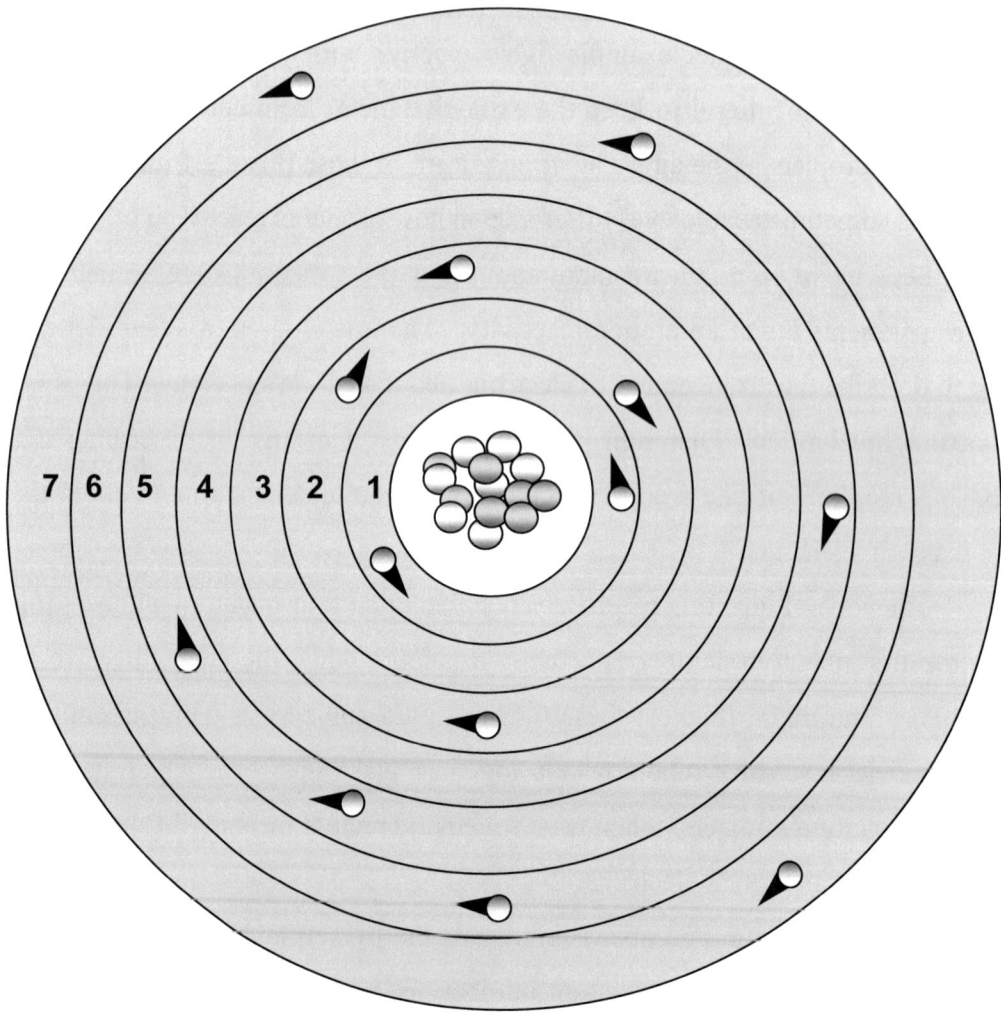

The principle quantum number indicates the relative distance an orbit or energy level is from the nucleus of an atom. Note the orbits numbered 1-7 with 1 being the closest to the nucleus and 7 being the farthest from the nucleus.

of these orbits (or energy levels) as being like the layers of rubber bands you find inside a golf ball. However, not all of these orbits are spherical or ball-shaped.

As we mentioned above, not all of the paths that the electrons follow are spherical or ball-shaped. The second quantum number, known as the **orbital quantum number**, allows us to indicate the shape of the orbit or path that the electrons follow. There are four possible path shapes that the electrons may follow. However, only two of those shapes are readily described: spherical and pear-shaped. We indicate these shapes by using a lower-case "s" to indicate spherical and a "p" to indicate pear-shaped. The two remaining shapes, although not described by chemists, are given the letters "d" and "f."

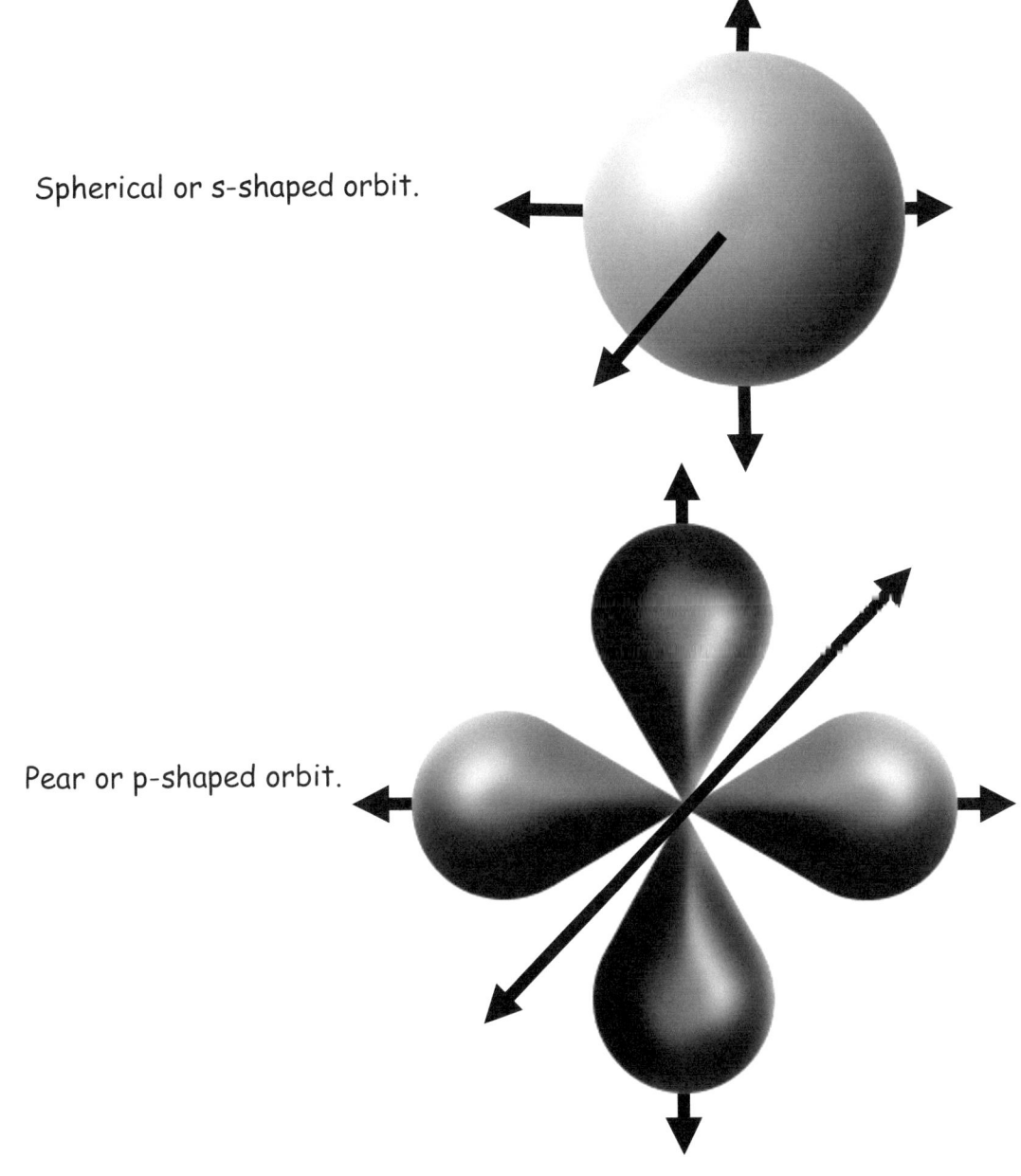

Spherical or s-shaped orbit.

Pear or p-shaped orbit.

Some chemists feel the d-shaped orbit is shaped like a dumbbell. To review, the second quantum number (the orbital quantum number) is a letter (s, p, d or f) which tells the shape of the orbit that the electrons follow.

There can be more than one shape of orbit within each orbit or energy level. In other words, you might have a combination of shapes within each orbit or energy level. It just so happens that the principle quantum number equals the variety of shapes within each energy level. For example, the first energy level, principle quantum number of 1, has only 1 shape of orbit (which is spherical). The second energy level, principle quantum number of 2, has two shapes within the level: spherical and pear-shaped. Can you guess how many shapes are found within the third energy level? You're correct if you said 3! Those shapes would be spherical, pear-shaped and the shape indicated by the letter d. The sequence of shapes is spherical (s), then pear-shaped (p), then the d shape and, finally, in the fourth energy level, the f shape. We will primarily be using the elements whose electrons fill the s-, p- and d-shaped orbitals.

Well, we're half-way through this discussion of how electrons are arranged in each atom and how the arrangement of electrons around the nucleus of any atom determines the reactivity of that atom. Let's pause and review what we talked about so far. We've discussed the principle quantum number which tells the position or distance an orbit (or energy level) is away from the nucleus (whole numbers, 1-7, 1 being the closest and 7 being the farthest from the nucleus). Next, we discussed the orbital quantum number which indicates the shape of the orbit (letters s, p, d or f; s being spherical, p being pear-shaped). If you feel at all confused at this point, consider re-reading the chapter up to this point. As with most things, as you apply the concepts outlined above, you will undoubtedly feel more comfortable with them.

The third quantum number, **the magnetic quantum number**, tells the direction or orientation in space that each orbit has. The magnetic quantum number tells how the orbit is positioned along imaginary lines or axes which go right through the center of the nucleus of the atom. These axes are given the notations of x, y and z. The x axis runs right and left, the y axis runs up and down, and the z axis runs *into* and *out of* the page. If you laid a pencil across this page, it would be aligned with the x axis. If you laid a pencil going up-and-down on this page, it would be aligned with the y axis. If you *in-*

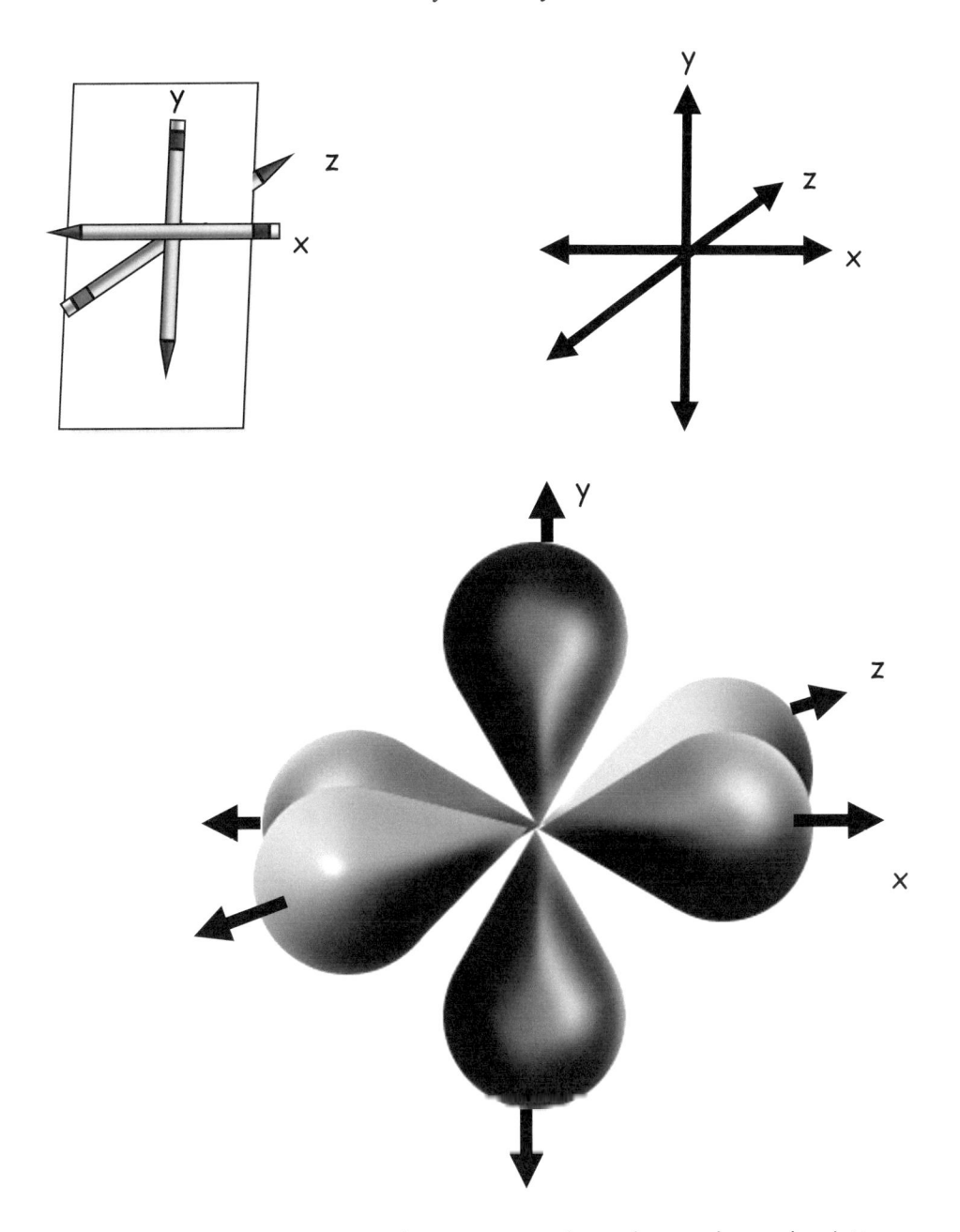

The magnetic quantum number states that the p-shaped orbits may be aligned along three imaginary axes: x, y and z.

serted a pencil *through* this page, it would be aligned with the z axis. The magnetic quantum number utilizes these three orientations about the center of an atom: x, y and z. It is only the p or pear-shaped orbits that we are concerned with regarding the x, y and z orientations. Look at the diagrams on the preceding page to see these three orientations of the pear-shaped orbits.

The fourth quantum number is the simplest of all to visualize. This quantum number, known as the **spin quantum number**, tells the direction the electrons are rotating, hence the name spin quantum number. Not only are the electrons traveling in an orbit around the nucleus in an atom, they are also spinning much like the earth rotates on its axis as it revolves around the sun. The direction of spin can either be clockwise or counter-clockwise. It is theorized that two electrons can be found in each shaped orbit, i.e. 2 electrons in each s-shaped orbital, 2 in each p-shaped orbit and so on. One of these electrons is thought to be spinning clockwise while the other spins counter-clockwise.

The spin quantum number says that the electrons travel in pairs with one electron spinning clockwise and the other spinning counter-clockwise. Elements with an odd number of electrons will have one lone electron in its outermost orbit.

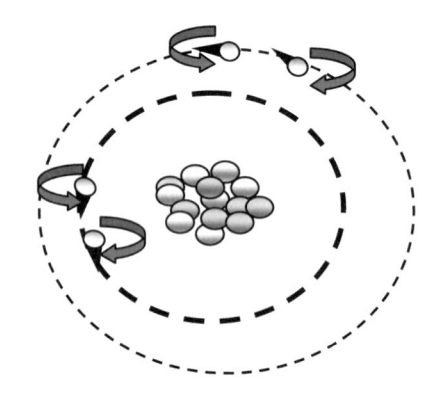

Quantum Number	And what that Quantum Number Means
Principle Quantum Number	Tells the position or **layer** (energy level) that the electrons are traveling in. Whole numbers: 1-7. 1 is closest to the nucleus.
Orbital Quantum Number	Tells the **shape** of the path in which the electrons are traveling (s, p, d and f).
Magnetic Quantum Number	Tells the **orientation** in space of the p-shaped orbits along the x, y and z axes.
Spin Quantum Number	Tells that within each **pair** of electrons, one electron **spins** clockwise and the other spins counter-clockwise.

Name_____ Date_____

Friendly Chemistry

Quantum Mechanics Review

Fill in the blanks with the appropriate words. Refer to your text if you need help!

    In our discussion of quantum mechanics, we said there were _____ quantum numbers. The first quantum number is known as the _____ and it said that _____ are arranged in _____. It said that there can be up to _____ layers. The layer closest to the nucleus is the _____ layer. The second quantum number is known as the _____. It says that electrons are found in different _____ paths as they move around the nucleus of an atom. There are _____ different shapes of paths: _____, _____, _____ and _____. The letter abbreviations for these shapes are: _____, _____, _____ and _____.

    The third quantum is called the _____. It says that the _____-shaped orbits can be found in _____ different orientations in space. These three orientations are labeled as _____, _____ and _____.

    The fourth quantum number is called the _____. It says that _____ travel in _____ about the nucleus of the atom. One electron will _____ _____ while the other electron will _____ _____.

    It is important to know about quantum mechanics because quantum mechanics will _____ why some elements are very _____ while others are very _____. By knowing how the electrons are arranged in various atoms helps one _____ _____.

S35

Notes

Lesson 4: Applying Quantum Mechanics

In the last lesson we discussed a code known as **quantum mechanics** that would enable us to describe how electrons are arranged about the nucleus of a particular atom. We stated that knowing this arrangement would enable us to understand the nature of those atoms, i.e. how stable they were or how reactive they were with other atoms.

In order to learn the application of the code to our model of the atom, we will use a study aid known as the Doo-Wop Board. In your box of manipulatives which came with this book, find a large card labeled Doo-Wop Board. You will also find a bag of colored plastic chips. Choose two different colors of chips and separate them from the rest of the chips. You will need approximately 20 chips of each color.

Across the bottom of the Doo-Wop Board™ you will see the word "nucleus" written which indicates that everything on the board above the term "nucleus" will be considered outside the nucleus (in the electron cloud). The Doo-Wop Board™ represents a section of the electron cloud. Look at the diagram on the next page.

Friendly Chemistry

7 6 5 4 3 2 1

Orbit/layer 5
Orbit/layer 4
Orbit/layer 3
Orbit/layer 2
Orbit/layer 1
Nucleus

4d
5s
4p
3d
4s
3p
3s
2p
2s
1s

Doo-Wop Board

NUCLEUS

The Doo-Wop Board is analogous to a section of the electron cloud.

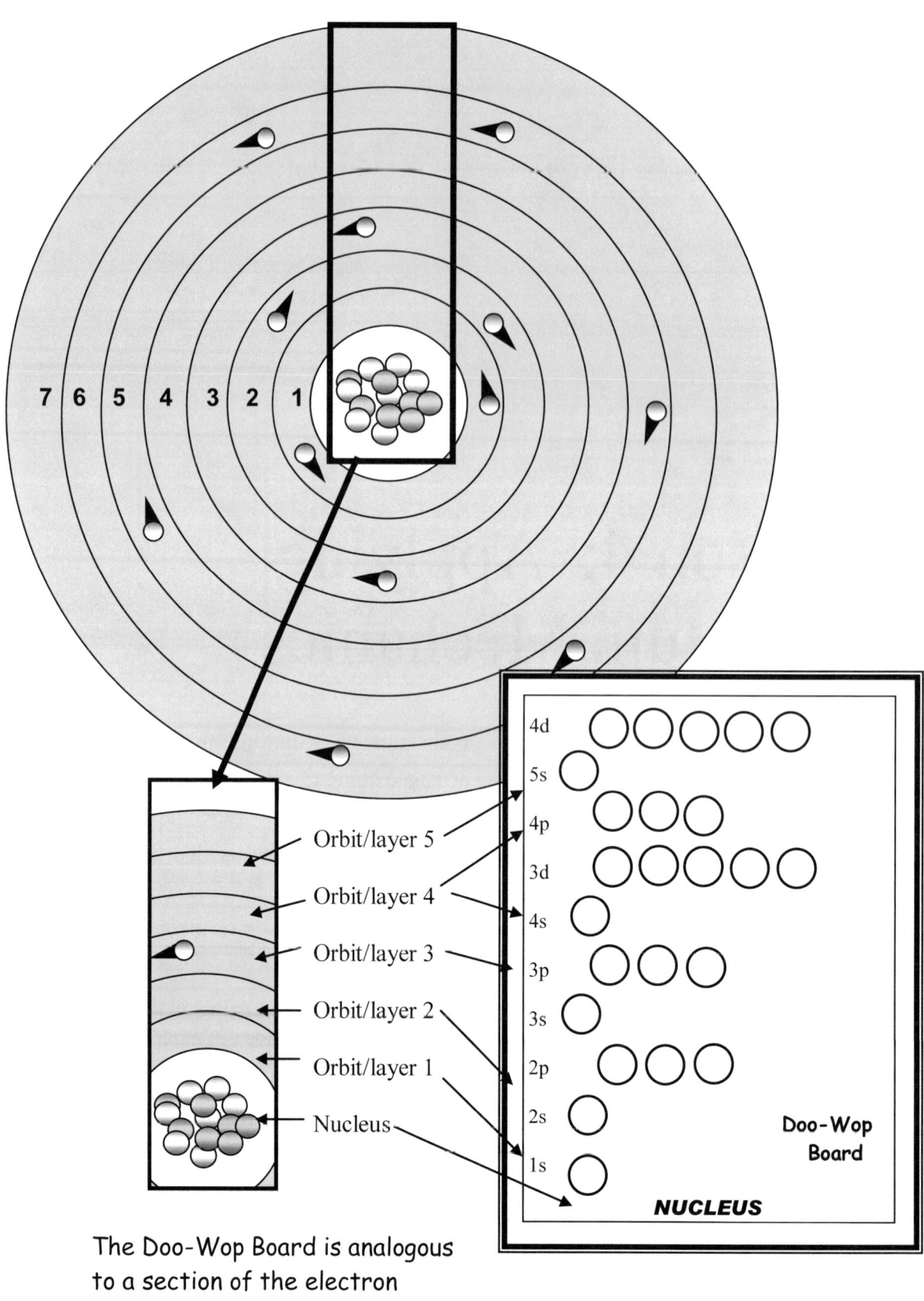

Look at the first circle just above the nucleus. To the left of the circle you will see the notation 1s. The 1s indicates that the circle will hold the electrons in the first orbit or shell (1) from the nucleus which happens to be spherical in shape (s). Each circle above the 1s circle is another orbit of particular shape. The sequence that the circles are placed on the board is significant, but we will discuss that later. At this point, realizing that each circle represents an orbit of a particular shape is important.

The doo-wops (chips) that you have separated into two piles represent **electrons**. These electrons will be placed into the circles to represent the electrons in the atom. Let's look at an example!

Example 1: Let's start with an atom of helium. The first thing that you will need to find out is how many electrons will be found in an atom of helium. Do you recall from a previous chapter how to locate this information? Yes, the periodic table of elements (that chart with all the element symbols and names) has within each element block the atomic number which tells us the number of electrons or protons or, generally, the number of neutrons found in an atom of that particular element. You probably will recall that the symbol for helium is He and can be found over in the upper right hand corner of the table. There you will see that the atomic number for helium is 2. This means that an atom of helium will have 2 electrons, 2 protons, and generally, 2 neutrons. The number of electrons (2) is important to us right now. See the diagram below help you locate helium.

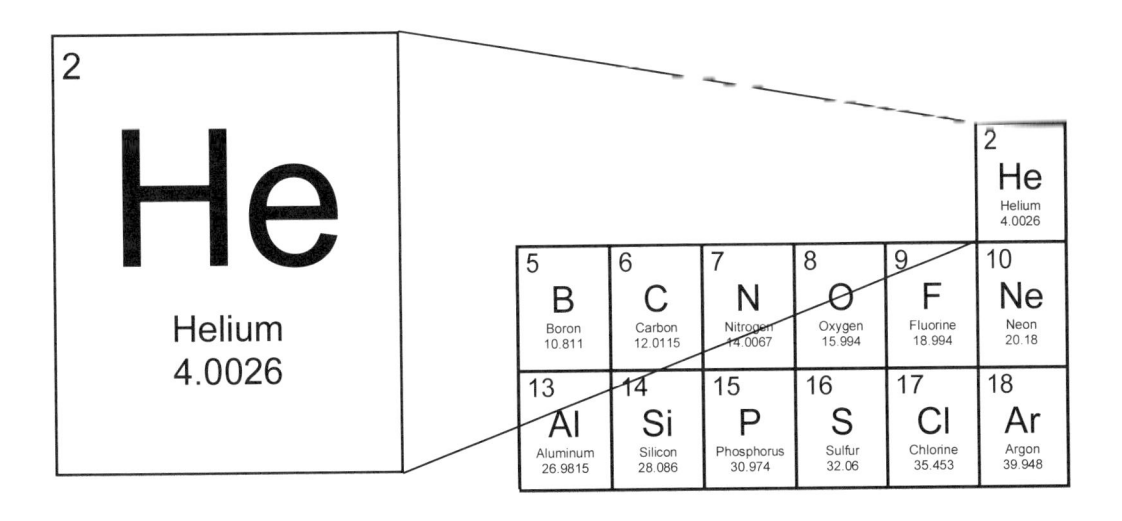

Now, take two of your doo-wops (electrons), one of each color. The first rule of how electrons fill the orbits is that electrons begin filling the orbits nearest the nucleus first and proceed outward (upward on the Doo-Wop Board). Imagine that you are holding your Doo-Wop Board in a sink. As you fill the imaginary sink with water, the bottom cup or orbit will fill first. That cup is the 1s orbit. Then the 2s orbit will fill and then the 2p orbits and so on. This is exactly how you will place your doo-wops (electrons) into the cups: from the bottom up (or to be more precise, from the inside of the atom outward).

Now, let's get back to our example of helium. You found from the periodic table of elements that each atom of helium has two electrons traveling in its electron cloud. Following the "fill rule" above, you should place the two doo-wops you are holding into the first (bottom) circle. Recall that within each shape of orbit you will find two electrons, hence we placed both of the electrons that an atom of helium has in the first orbit. Look at the diagram below to see what your Doo-Wop Board should look like. Note that we placed one chip of each color in the 1s orbit.

You may be wondering why you are using two colors of doo-wops. The answer is that the two colors of doo-wops represent the two possible spins of those electrons: clockwise and counter-clockwise (the spin quantum number).

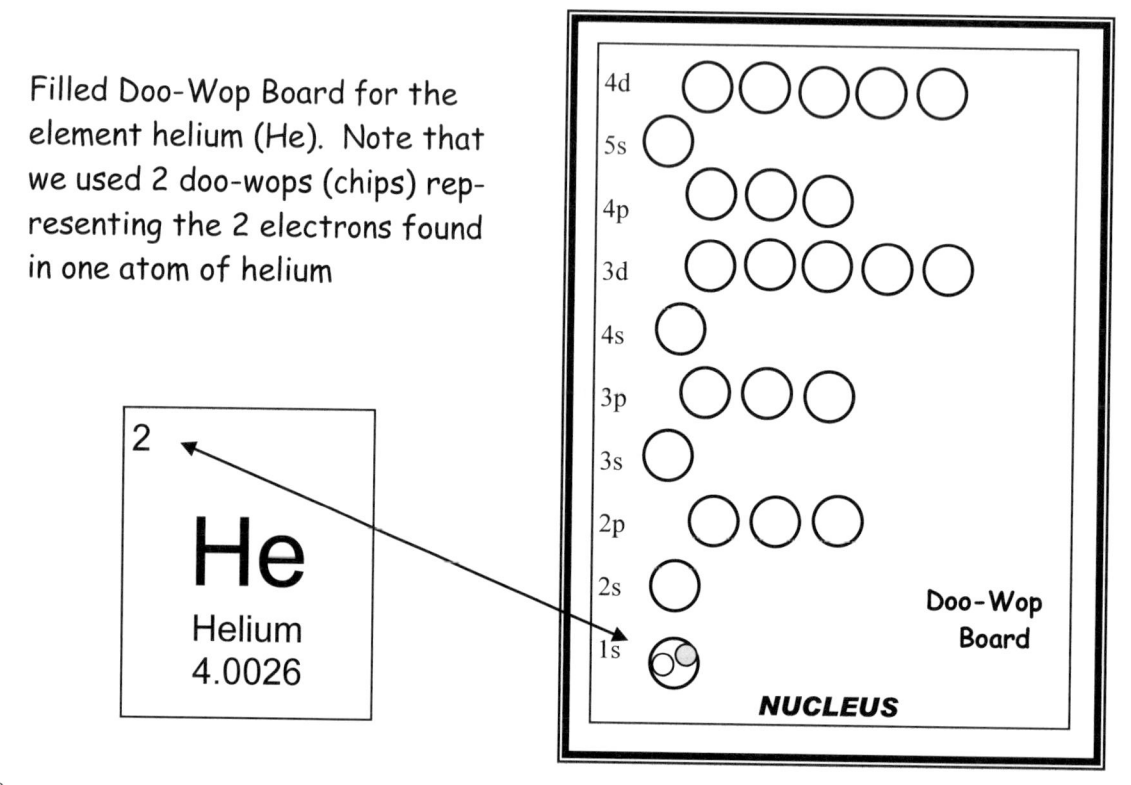

Filled Doo-Wop Board for the element helium (He). Note that we used 2 doo-wops (chips) representing the 2 electrons found in one atom of helium

2

He
Helium
4.0026

Friendly Chemistry

By looking at your Doo-Wop Board, you can describe helium as having two electrons in its electron cloud and they are both found traveling in a spherical-shaped orbit on the first energy level from the nucleus. This is called "reading" your board.

At this point in our discussion, we haven't yet explored what this information means regarding the reactivity or stability of helium. You will find a considerable amount of material in the next lessons which will help you make inferences regarding the reactivity or stability of each element. For now, let's look at another example.

Example 2: Describe the arrangement of electrons found in an atom of carbon. Find carbon on your periodic table of elements. The symbol for carbon is C. If you look on the right-hand side of the table you will see the element carbon on the first row of elements going across the top of the table.

The atomic number for carbon is 6 meaning that an atom of carbon has 6 electrons, 6 protons and, generally, 6 neutrons. We are concerned with the number of electrons which is 6. Take 6 doo-wops, representing the 6 electrons, three of each of the

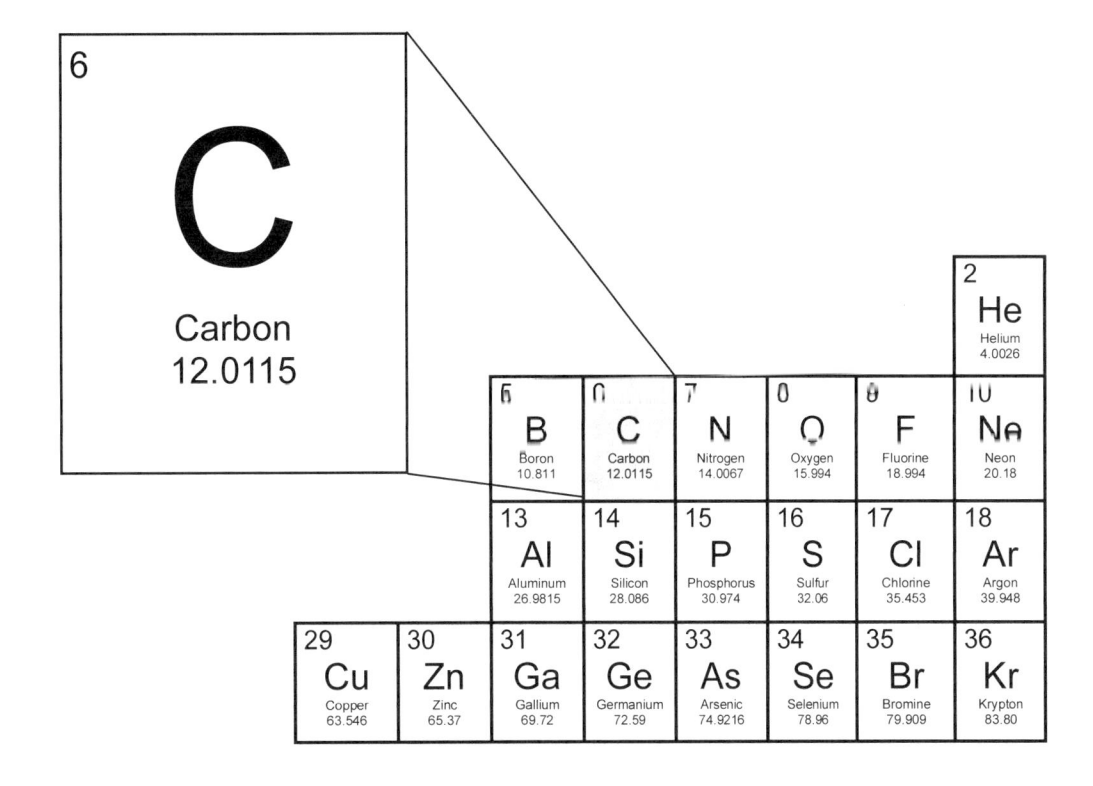

S41

two colors. Using the first fill rule, place two electrons in the 1s cup, then two more electrons in the next cup up (the 2s cup). Now you come to a series of three cups at the 2p level. Remember in Lesson 3 when we discussed the magnetic quantum number (the quantum number which tells us the orientation in space in which the orbits lie)? We stated that the pear-shaped orbits (p) were the only shaped orbits for which knowing an orientation in space would be helpful when describing the arrangement of electrons in the electron cloud.

Now, look back at your Doo-Wop Board at the 2p level and beneath each of the three circles you will see an x, y or z. These notations represent each of the three possible orientations in space in which you might find electrons traveling in the pear-shaped orbits. Refer to Lesson 3 if you would like to review the concepts we discussed regarding the magnetic quantum number.

Let's return to our example of carbon. We have used four electrons so far (in the 1s and 2s cups) and have two remaining. The second fill rule states that one electron is placed in each orientation position (x, y and z) and if necessary, a second is placed in each position. In other words, you place one of your doo-wops in the $2p_x$ circle, a second doo-wop in the $2p_y$ circle and, if necessary, one more doo-wop in the $2p_z$ circle. Then, if necessary, you go back and place a second doo-wop in each circle. The diagram on this page shows you the sequence of filling the p orbits.

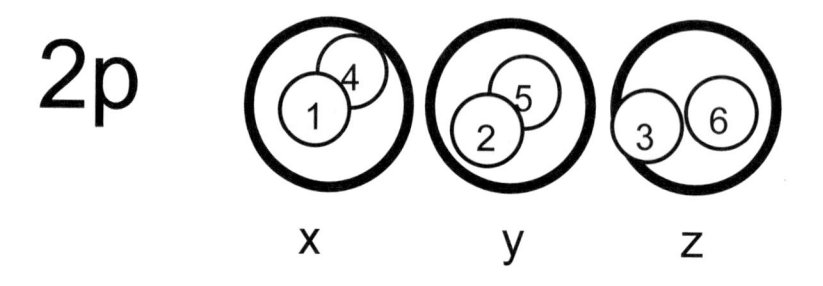

In the case of carbon, we have two remaining doo-wops to place on the board. Place one doo-wop in the $2p_x$ circle and the last doo-wop in the $2p_y$ circle. Note that it does not matter which color you place in the circle first, as long as you have one doo-wop of each color to represent the clockwise and counter-clockwise spins of the electrons. Look at the diagram below to see how you should have filled your Doo-Wop Board to represent the arrangement of electrons for carbon.

Friendly Chemistry

2p ⬤◯◯
    x y z

Note that when filling electrons on the pear-shaped orbits as in this example of carbon, one electron goes into each direction orbit before a second electron fills the circle.

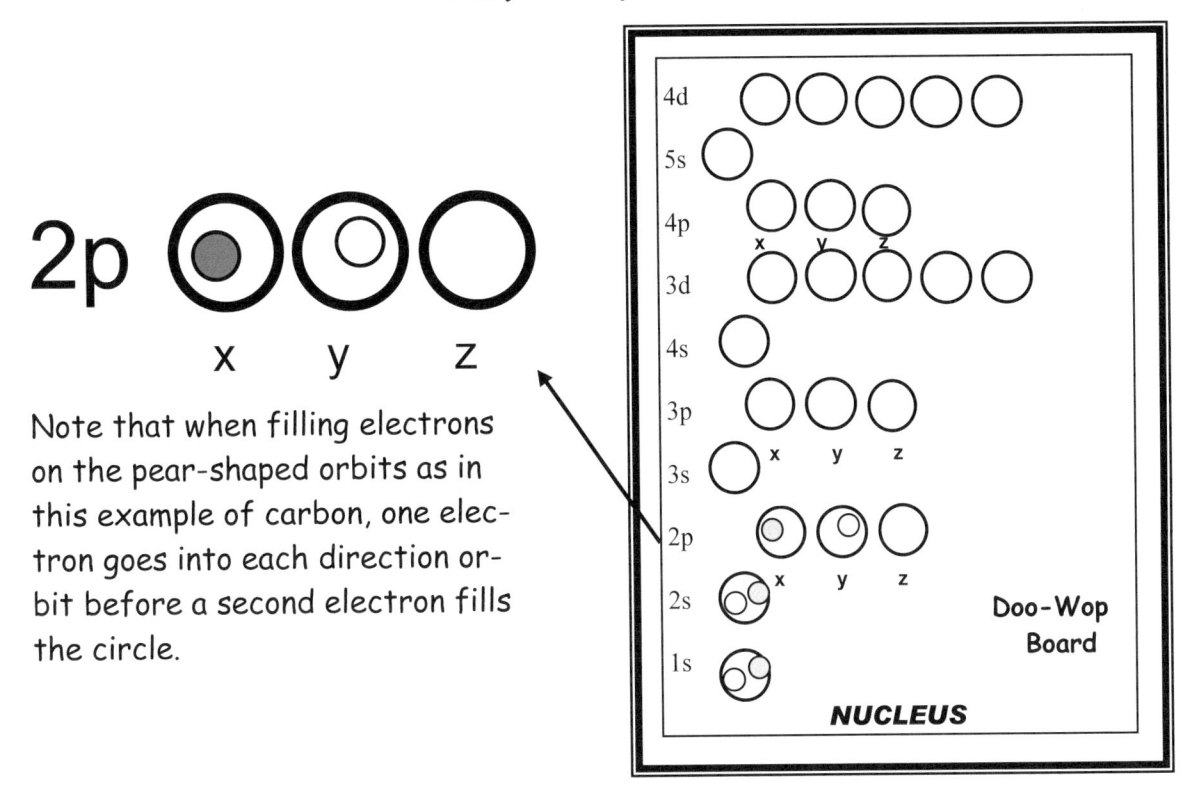

Doo-Wop Board

NUCLEUS

To complete this example, let's make a statement regarding the arrangement of the electrons found in an atom of carbon. We can say by looking at our Doo-Wop Board™ that each atom of carbon contains six electrons: two electrons are found in the spherical-shaped orbit on the first energy level, two are found in the spherical-shaped orbit on the second energy level and two are found in the pear-shaped orbits - one in the x orientation and one in the y orientation - also on the second energy level.

Example 3: Describe the arrangement of electrons found in an atom of magnesium. Find magnesium on your periodic table of elements. Its symbol is Mg (do not get confused with manganese whose symbol is Mn!) and can be found on the left side of the table in the second column from the left. See the diagram below to assist you in finding magnesium on the periodic table of elements.

S43

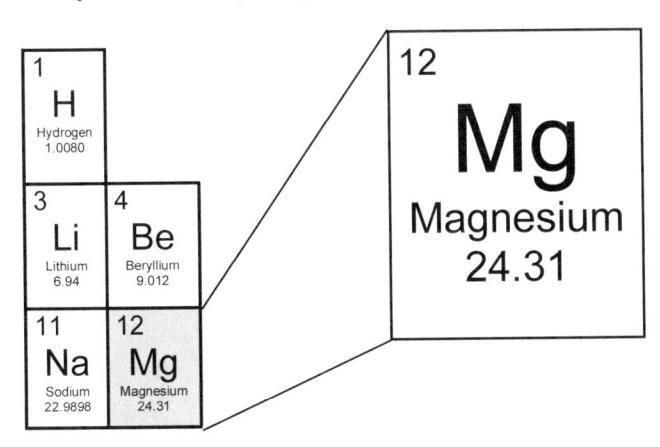

You should have two electrons in the 1s orbit, two electrons in the 2s orbit, and then by following the second fill rule, you should place one electron in the $2p_x$ orbit, one in the $2p_y$ orbit, one in the $2p_z$ orbit and then return to place a second electron (of different spin/color) in the $2p_x$ orbit, a second electron in the $2p_y$ orbit and a third in the $2p_z$ orbit. Were you able to follow that sequence? Look at the diagram below to help you follow the sequence of placing the electrons in the p orbits.

You should still have two electrons left in your hand. They will fill the next orbit which is the 3s orbit. Look below to see how your completed arrangement of electrons in an atom of magnesium should appear on your Doo-Wop Board.

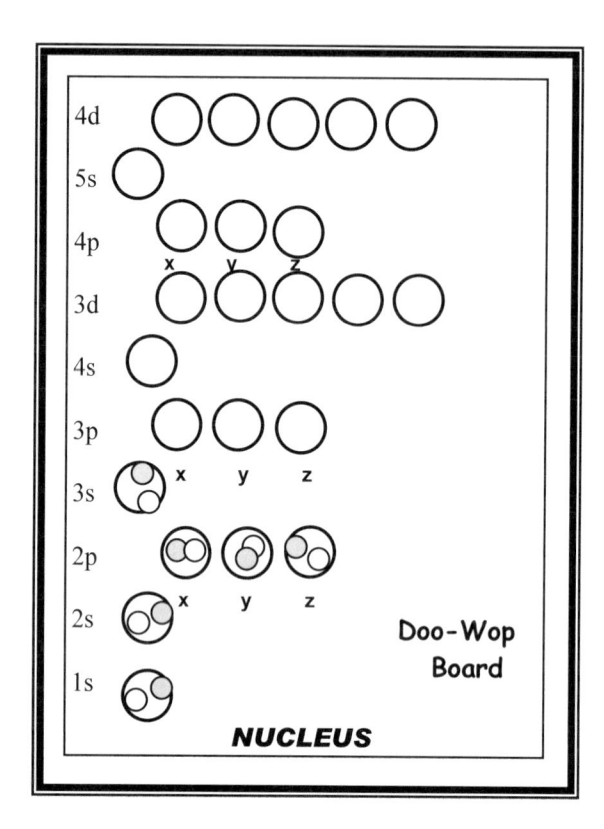

Filled Doo-Wop Board for the element magnesium (Mg). Note that a total of 12 doo-wops were used representing the 12 electrons in an atom of magnesium.

Based upon your Doo-Wop Board you can "read" your board and say that magnesium has 12 electrons in its electron cloud with two in the spherical shaped s orbit on the first energy level, two electrons in the s orbit on the second energy level, a total of six electrons divided among the three p orbits (x, y and z) and two electrons in the s orbit at the third energy level.

For more practice using your Doo-Wop Board, you will now play the Doo-Wop Board Filling version of Teamwork!

Lesson 5: Learning the Code: Orbital Notation

In Lesson 4 we discussed how using the Doo-Wop Board can enable you to describe the arrangement of electrons in an atom of a particular element. You learned to first check the periodic table of elements to find the element's atomic number which indicates the number of electrons in an atom of that element. Then you proceeded to fill your Doo-Wop Board with doo-wops representing the number of electrons found in that atom. Based upon which orbits (circles) you filled, a statement could then be made which described the arrangement of electrons in an atom of that particular element.

The statement that was made, although accurate, was generally very lengthy and sometimes rather complex. This chapter was written to help you take full advantage of the quantum mechanics "code" that you were introduced to in the last lesson. You will learn how to take the orbit notations (the 1s's and the 2p's, etc.) and join them together in a readable, usable system that others, who also know the code, can understand. There are three systems, or notations, which are used. We will introduce the first notation, known as **orbital notation**, and then, using it as our standard, we will introduce the second notation, known as **electron configuration notation**. The third notation, known as the **electron dot notation**, is slightly more complex, but continues to build upon concepts learned in the first two notations. So, get out your Doo-Wop Board and doo-wops and let's get started!

Orbital Notation

The orbital notation is straightforward and easy-to-learn if you have your Doo-Wop Board beside you. In this notation, you will use small **arrows** to symbolize electrons. Probably the best way to learn this notation is to look at an example.

Example 1: Describe the arrangement of electrons in an atom of <u>carbon</u> using *orbital notation*.

First, just as you did in the last lesson, find carbon on the periodic table of elements. Then fill your Doo-Wop Board. Look at the diagram here to check your work. Remember to use three doo-wops of one color and the rest the other color to represent the two spins of the electrons.

As we stated above, orbital notation uses small arrows to symbolize electrons. The key to orbital notation is that you should use as many arrows in your notation as you have electrons in your atom (or doo-wops on your Doo-Wop Board). In our example of carbon, we found that we used six doo-wops which represented the six electrons found in an atom of carbon. We began at the bottom of our Doo-Wop Board and filled each orbit with two doo-wops (one of each color to represent the clockwise and counter-clockwise spins of the electrons). Based upon that filling, we can now write our orbital notation to describe the electrons in an atom of carbon as follows:

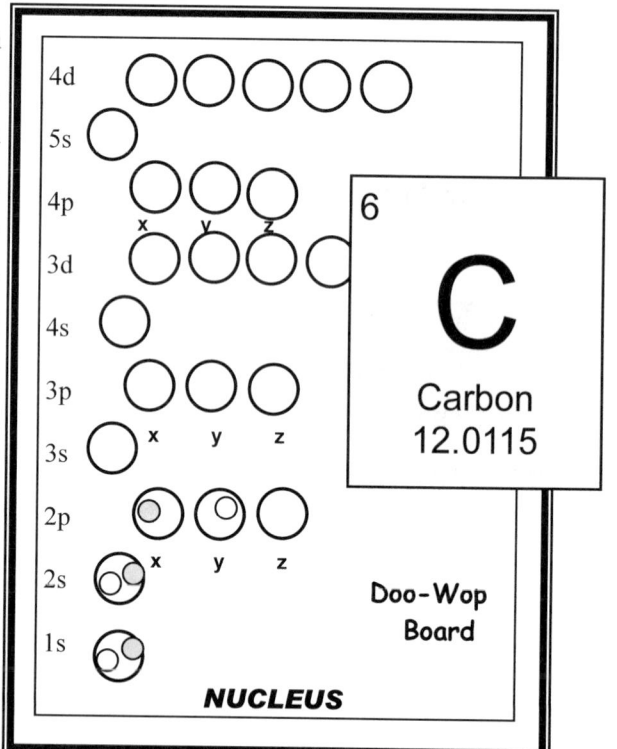

$$\text{C:} \quad \underline{\updownarrow}_{1s} \quad \underline{\updownarrow}_{2s} \quad \underline{\uparrow}_x \underline{\uparrow}_y \underline{}_z \atop 2p$$

S46

Note that we first write the symbol of the element we are describing, followed by a colon.

C:

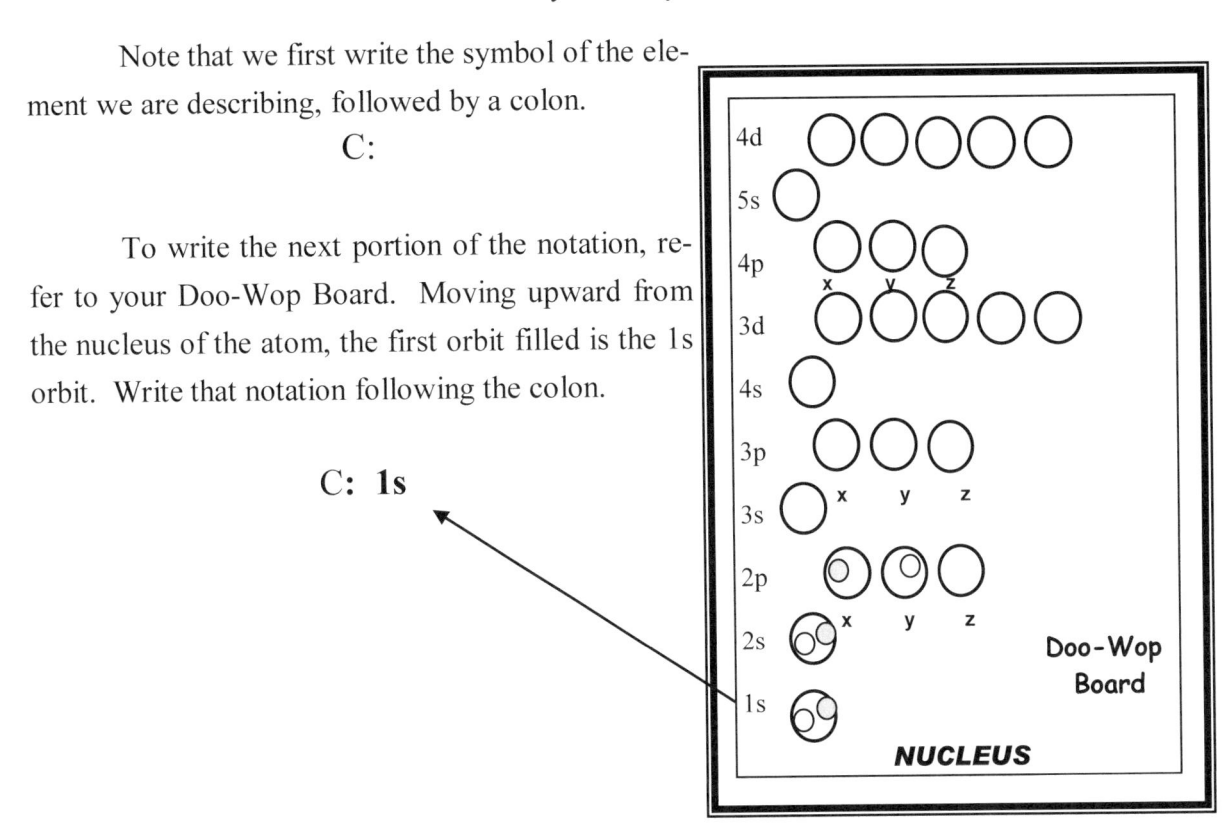

To write the next portion of the notation, refer to your Doo-Wop Board. Moving upward from the nucleus of the atom, the first orbit filled is the 1s orbit. Write that notation following the colon.

C: 1s

Above the 1s notation, draw a line and above that line draw two arrows representing the two doo-wops (electrons) found in the 1s orbit. Note that we draw one arrow pointing upward and one downward to represent the clockwise and counterclockwise spins of the two electrons found in that orbit.

$$\underline{\downarrow\uparrow}$$
C: 1s

Next, refer back to your Doo-Wop Board. Moving upward from the 1s orbit, the next orbit filled by the electrons in an atom of carbon is the 2s orbit. Continue the notation by writing 2s and two arrows over a line above the 2s notation.

$$\underline{\downarrow\uparrow} \quad \underline{\downarrow\uparrow}$$
C: 1s   2s

Refer again to your Doo-Wop Board. Moving upward from the 2s orbit you will find that the next set of orbits filled is the 2p orbits. Note that you have placed one doo-wop in the $2p_x$ orbit and one doo-wop in the $2p_y$ orbit. The $2p_z$ orbit has no doo-wops. In order to represent the locations and orientations of these electrons using the

S47

orbital notation, write the 2p notation with three lines above it: the first correlating with the x orientation, the second correlating with the y orientation and the third line correlating with the z orientation.

$$\text{C:} \quad \underset{1s}{\uparrow\downarrow} \quad \underset{2s}{\uparrow\downarrow} \quad \underset{{}^x{}^y2p{}^z}{\underline{}\,\underline{}\,\underline{}}$$

Complete the orbital notation by drawing in one arrow for each doo-wop present in the 2p$_x$ orbit and the 2p$_y$ orbit. Note that the 2p$_z$ will remain empty in the notation.

$$\text{C:} \quad \underset{1s}{\uparrow\downarrow} \quad \underset{2s}{\uparrow\downarrow} \quad \underset{{}^x{}^y2p{}^z}{\underline{\uparrow}\,\underline{\uparrow}\,\underline{}}$$

To check your work, count the number of arrows drawn in the notation. The number of arrows should equal the number of doo-wops on your Doo-Wop Board™ (which equals the number of electrons found in an atom of carbon: six).

Let's try another example.

Example 2: Describe the arrangement of electrons in an atom of <u>magnesium</u> using *orbital notation*.

Begin by finding magnesium on the periodic table of elements. The symbol for magnesium is Mg. The atomic number for magnesium is 12, indicating that each atom of magnesium has 12 electrons, 12 protons and, generally, 12 neutrons.

Begin filling your Doo-Wop Board™ with 12 doo-wops (remember to use six of each color). Check your work by referring to the diagram here.

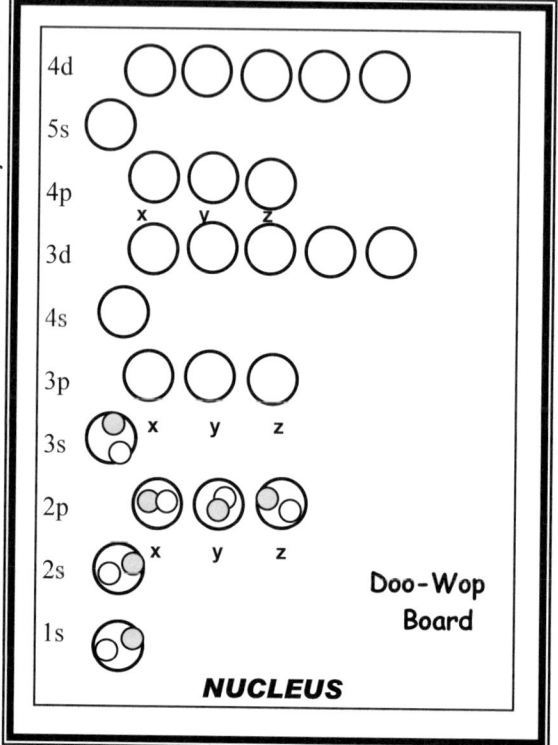

S48

Begin the notation by writing the symbol for magnesium followed by a colon.

Mg:

Then, as we proceeded in the example above, begin with the first orbit filled just outside the nucleus (1s). In an atom of magnesium, this orbit has two electrons (refer to your Doo-Wop Board). Indicate these two electrons by drawing two arrows above the 1s notation. Remember to draw one arrow going upward and one downward representing the spins of the electrons.

$$\underline{\uparrow\downarrow}$$
Mg: **1s**

Continue with the next orbit filled which, in the case of magnesium, is the 2s orbit. Write that notation along with the two arrows representing the two electrons found in that orbit.

$$\underline{\uparrow\downarrow} \quad \underline{\uparrow\downarrow}$$
Mg: 1s   **2s**

Continue with the next orbit which would be the 2p orbits. In the case of magnesium, each of the 2p orbits (x, y and z) is filled with electrons. As we did in the example of carbon, represent those filled orbits by writing the 2p notation with three locations above for each orbit. Draw two arrows in each location representing the two arrows in each orbit.

$$\underline{\uparrow\downarrow} \quad \underline{\uparrow\downarrow} \quad \underline{\uparrow\downarrow} \; \underline{\uparrow\downarrow} \; \underline{\uparrow\downarrow}$$
Mg: 1s   2s $^x \quad ^y \mathbf{2p} \; ^z$

Complete the notation for magnesium by referring back to your Doo-Wop Board. Note that two electrons remain in the 3s orbit. Complete your notation as follows:

$$\underline{\uparrow\downarrow} \quad \underline{\uparrow\downarrow} \quad \underline{\uparrow\downarrow} \; \underline{\uparrow\downarrow} \; \underline{\uparrow\downarrow} \quad \underline{\uparrow\downarrow}$$
Mg: 1s   2s $^x \quad ^y 2p \; ^z$ **3s**

Check your work by counting the number of arrows you drew in the notation. If you counted twelve, you are correct! Remember that the number of arrows must equal the number of electrons found in an atom of the element you are describing. Practice what you've learned in this lesson by completing the practice pages found on the next pages of your text.

Name_____ Date_____

Friendly Chemistry

Orbital Notation Practice 1

Write the orbital notation for each element listed below. Remember to begin with the element symbol, followed by a colon.

Element	Orbital Notation
1. aluminum	Al:
2. potassium	
3. scandium	
4. phosphorus	
5. calcium	
6. lithium	
7. neon	
8. hydrogen	
9. sodium	
10. chlorine	
11. boron	
12. beryllium	
13. sulfur	
14. argon	

Friendly Chemistry

Name_____ Date_____

Friendly Chemistry

Orbital Notation Practice 2

Write the orbital notation for each element listed below. Remember to begin with the element symbol, followed by a colon.

Element	Orbital Notation
1. Carbon	C:
2. Sulfur	
3. Neon	
4. Lithium	
5. Nitrogen	
6. Magnesium	
7. Hydrogen	
8. Boron	
9. Oxygen	
10. Potassium	
11. Phosphorus	
12. Argon	
13. Calcium	
14. Helium	

Lesson 6: Electronic Configuration Notation

As we mentioned earlier, the electronic configuration notation (which is the second of three notations you will learn) builds upon the skills you acquired in learning the orbital notation. In orbital notation you used arrows to designate the electrons found in each orbit. In electronic configuration notation (ECN), you will replace those arrows with superscripts (small numbers written above the orbit notations, also known in math classes as "powers"). As we did with orbital notation, let's learn this notation by looking at some examples.

Example 1: Write the electronic configuration notation for an atom of <u>carbon</u>.

Begin solving this example problem by filling your Doo-Wop board to represent the electrons found in an atom of carbon. Recall that the symbol for carbon is C and the atomic number for carbon is six. Check the diagram below to see that you properly filled your Doo-Wop board. Remember that you can also check your work by counting

Friendly Chemistry

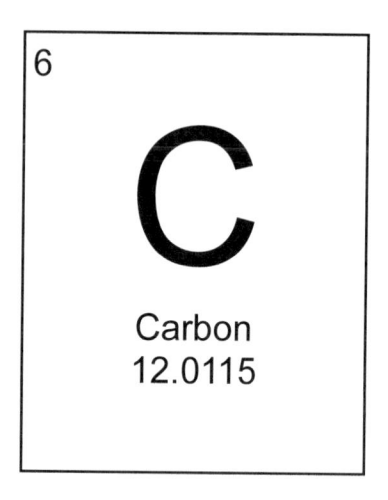

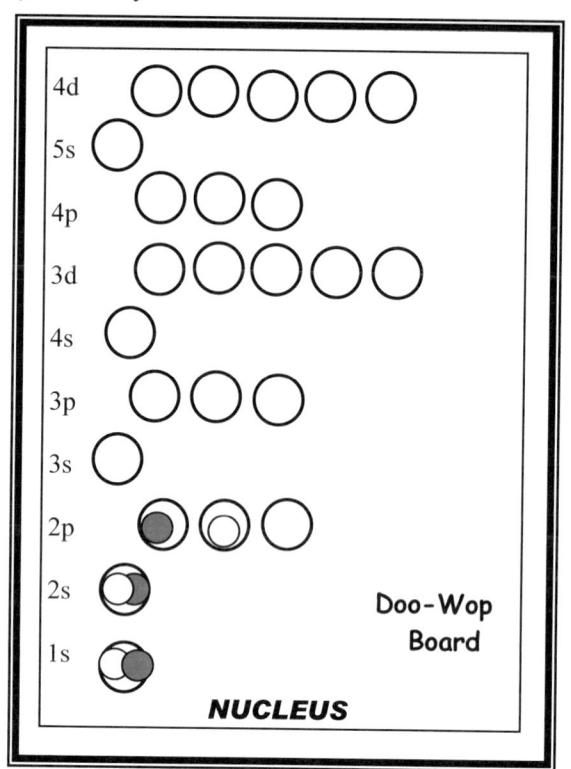

the number of doo-wops which, in the case of carbon, should equal six.

As we started with orbital notation, begin writing the electronic configuration notation by writing down the element symbol followed by a colon.

C:

Refer to your Doo-Wop Board and note that the first orbit filled outside the nucleus of the atom is the 1s orbit. Write down the 1s notation following the colon and, instead of writing arrows to designate electrons, write a superscript representing the number of electrons found in the 1s orbit (in this example of carbon, that superscript will be 2).

C: $1s^2$

Continue the notation by referring again to your Doo-Wop Board. Note that the next filled orbit is the 2s orbit. Write that notation and include the superscript indicating the number of electrons found in the 2s orbit (2).

C: $1s^2\ 2s^2$

Complete the notation by writing the next orbit which is the 2p orbits. Unlike the orbital notation, we will not designate the specific locations of the x, y and z orbits. Instead, count the total number of electrons in the 2p orbits and write that number (2 in our example of carbon) as a superscript following the 2p notation.

$$C: 1s^2\ 2s^2 \mathbf{2p^2}$$

To check your work, simply add up the superscripts. The total should equal the total number of electrons found in an atom of that element. In our example (2+2+2=6), the sum of the superscripts is six which equals the total number of electrons found in an atom of carbon (6). Let's try another example.

Example 2: Write the electronic configuration notation for the electrons found in an atom of <u>magnesium</u>.

Begin your work by filling your Doo-Wop Board with the appropriate number of doo-wops. Refer to your periodic table of elements to find the element magnesium. Note that the atomic number for magnesium is 12 (meaning that an atom of magnesium will have 12 electrons in its electron cloud). See the diagram below to make sure you've correctly filled your board.

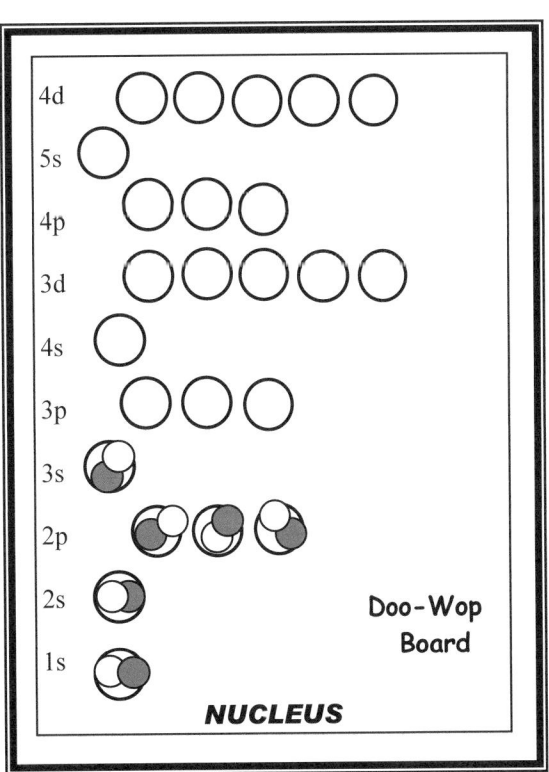

S55

Begin writing the electronic configuration notation for magnesium by writing the symbol for magnesium followed by a colon.

Mg:

Refer to your Doo-Wop Board to continue writing the notation. Write down the first orbit filled and, as a superscript, the number of electrons found in that orbit.

Mg: $1s^2$

Continue with the next orbit moving away from the nucleus.

Mg: $1s^2$ $2s^2$

Continue writing the notation by referring again to your Doo-Wop Board. As we stated in the first example, we will not designate the specific locations of the three 2p orbits. Instead, we will count the total number of electrons in the 2p orbits and write that total as a superscript following the 2p notation.

Mg: $1s^2$ $2s^2$ $2p^6$

To complete the electron configuration notation for magnesium, refer once again to your Doo-Wop Board. Note that two doo-wops (electrons) remain in the 3s orbit. Write down the 3s notation followed by the superscript of 2 to represent those two remaining electrons.

Mg: $1s^2$ $2s^2$ $2p^6$ $3s^2$

Check your work by adding up the superscripts (2+2+6+2=12). The total equals 12 which equals the atomic number (and number of electrons in an atom of magnesium). Like you did with orbital notation, you can still "read" the notation telling the location of all electrons in an atom of magnesium. Let's try one more example of an element we have not yet worked with.

Example 3: Write the electronic configuration notation for the electrons found in an atom of <u>argon</u>.

As in the two previous examples, first fill your Doo-Wop Board with the appropriate number of doo-wops to represent the electrons found in an atom of argon. Find argon on your periodic table of elements. Argon, whose symbol is Ar, can be found on the far right side of the table in the third series, or row, down. The atomic number for argon is 18 which means that you should use 18 doo-wops (9 of each color) to fill your Doo-Wop Board. Refer to diagram below to check your work.

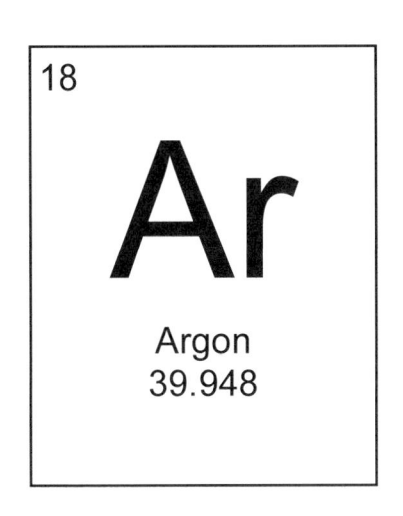

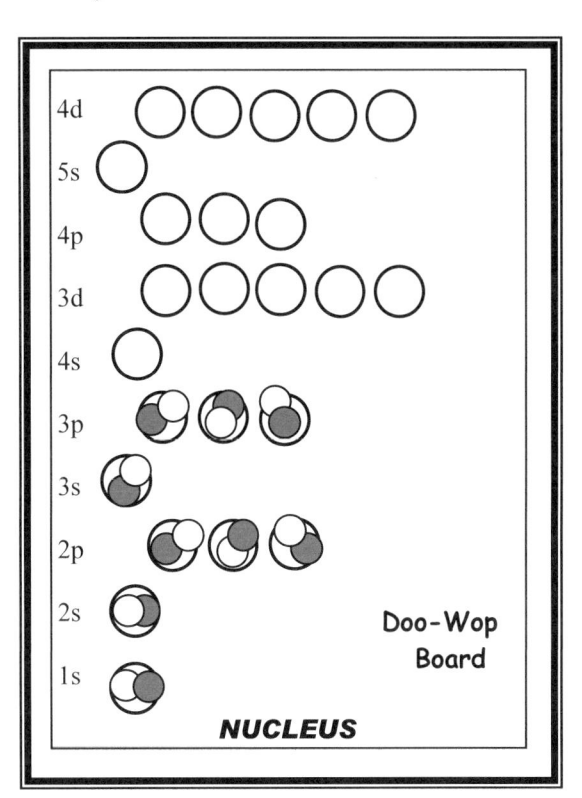

Begin writing the electron configuration notation for argon by referring to your Doo-Wop Board. Write down the symbol for argon followed by a colon.

Ar:

Continue the notation by writing the notation for the filled orbits as you move outward from the nucleus of the atom. The next two orbits you encounter are the 1s and 2s orbits.

Ar: $1s^2$ $2s^2$

Continue with the next two orbits you encounter.

$$\text{Ar: } 1s^2 \; 2s^2 \; \mathbf{2p^6 \; 3s^2}$$

Complete the notation by writing in the final orbit which contains electrons.

$$\text{Ar: } 1s^2 \; 2s^2 \; 2p^6 \; 3s^2 \; \mathbf{3p^6}$$

Remember to check your work by adding up the superscripts. In this example 2+2+6+2+6=18 which equals the number of electrons in an atom of argon. Like the two other ECN's (electronic configuration notations) we've completed, you can "read" the notation and tell the locations of all electrons in an atom of argon.

Continue to practice writing electronic configuration notations by playing *Doo-Wop Mania* Level 3 and the ECN version of the Teamwork Game. Your teacher will have instructions for play.

Name_____ Date_____

Friendly Chemistry

Electronic Configuration Notation Practice 1

Write the electronic configuration for each element listed below.

Element	Electronic Configuration Notation
1. Carbon	
2. Beryllium	
3. Oxygen	
4. Neon	
5. Nitrogen	
6. Boron	
7. Sodium	
8. Aluminum	
9. Sulfur	
10. Potassium	
11. Calcium	
12. Chromium	
13. Iron	
14. Fluorine	
15. Manganese	
16. Zinc	
17. Phosphorus	
18. Lithium	
19. Scandium	
20. Helium	

Friendly Chemistry

Name_____ Date_____

Friendly Chemistry

Electronic Configuration Notation Practice – 2

Write the electronic configuration for each element listed below.

Element	Electronic Configuration Notation
1. Nitrogen	
2. Oxygen	
3. Magnesium	
4. Silicon	
5. Fluorine	
6. Sulfur	
7. Neon	
8. Manganese	
9. Magnesium	
10. Strontium	

For each electron configuration notation written below, underline all of the errors. Tell the number of errors you found in the last column.

Element	ECN	Number of errors I found
11. Boron	Bo: $1s^2\ 2p^2\ 3p^1$	
12. Sodium	Na $1s^2\ 2s^2\ 2p^1\ 3s^5$	
13. Phosphorus	Ph: $1s^2\ 2s^2\ 2p^6\ 3s^3\ 4s^2$	
14. Potassium	K: $1s^2\ 2s^2\ 2p^6\ 3s^2\ 3p^6\ 4s^1$	
15. Calcium	CA: $1s^2\ 2s^2\ 2p^6\ 3s^2\ 3p^5\ 4s^2 3d^1$	

Lesson 7: Electron Dot Notation

In the last two lessons, you learned how to describe the arrangement of electrons using orbital notation (little arrows) and electronic configuration notation (using superscripts). In this lesson you will learn a third method of describing the arrangement of electrons which will bring you one step closer to being able to using electron arrangement to help you predict why some elements are very reactive while others are not. This third method is called electron dot notation.

Electron dot notation utilizes dots to represent those special electrons in an atom that are capable of interacting with other atoms and possibly forming bonds with those neighboring atoms. The special electrons are known as **valence electrons**. The valence electrons are those electrons that are in the outermost orbits or energy levels. Recall from our discussion of quantum numbers that the principle quantum number tells us the position or energy level on which we can find traveling electrons. We discussed that the higher the principle quantum number, the farther away from the nucleus the path or orbit in which the electrons were traveling. Therefore, the electrons that are found in the highest energy level (greatest principle quantum number) are considered the valence electrons.

Friendly Chemistry

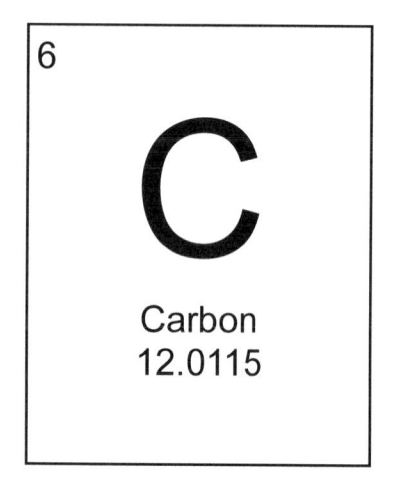

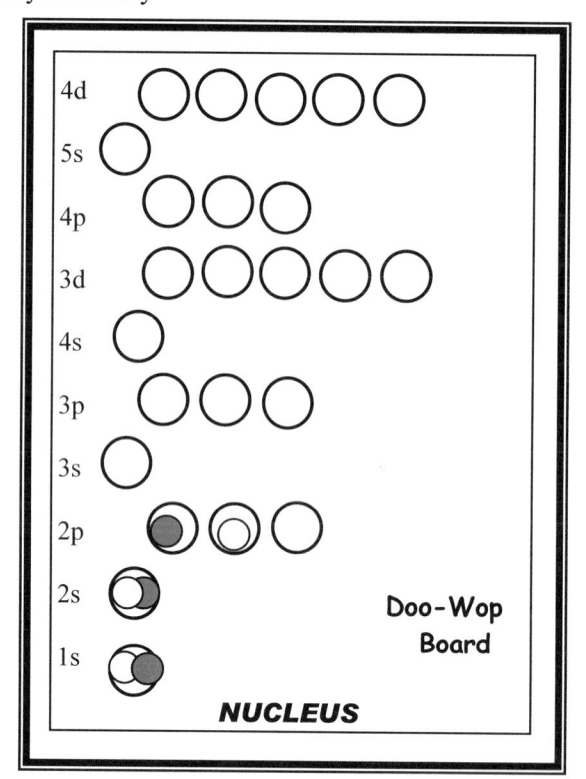

Let's look at an example of how to use electron dot notation to describe the arrangement of electrons about the nucleus of an atom.

Example 1: Use electron dot notation to describe the arrangement of electrons in an atom of <u>carbon</u>.

To begin writing electron dot notation, it is essential that you initially write the electron configuration notation (ECN) for the atom you are describing. In our example of carbon, the electron configuration notation for carbon is:

$$C:\ 1s^2\ 2s^2 2p^2$$

Refer to the Doo-Wop Board in the diagram above or go back to Lesson 6 if you are uncertain how to write the electron configuration notations for electrons of an atom.

The next step in writing the electron dot notation for carbon is to analyze the electronic configuration notation to find the number of valence electrons. Remember, the valence electrons are those electrons in the outermost orbits (energy levels). In the example of carbon, the outermost orbits filled are 2s and 2p (the highest principle quantum number is 2). We've drawn a box around those electrons in the notation below.

S62

C: 1s² |2s² 2p²|

To determine the number of valence electrons, add up the number of electrons in these outermost orbits. In the example of carbon, 2+2=4. This means that carbon has <u>four</u> valence electrons.

To write the electron dot notation for carbon, first write the symbol for carbon. (Note that no colon is necessary for electron dot notation!)

C

Represent each valence electron by making a dot for each valence electron around the element symbol. It is convention to place no more than two dots on any one side of the symbol and to evenly distribute the dots around the symbol as much as possible.

:C:

Try another example:

Example 2: Write the electron dot notation for <u>magnesium</u>.

Begin by writing the electronic configuration notation for magnesium. Recall that the atomic number for magnesium is 12 and that the electronic configuration notation for magnesium is:

Mg: 1s² 2s² 2p⁶ 3s²

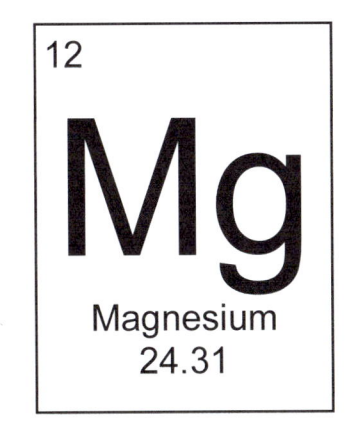

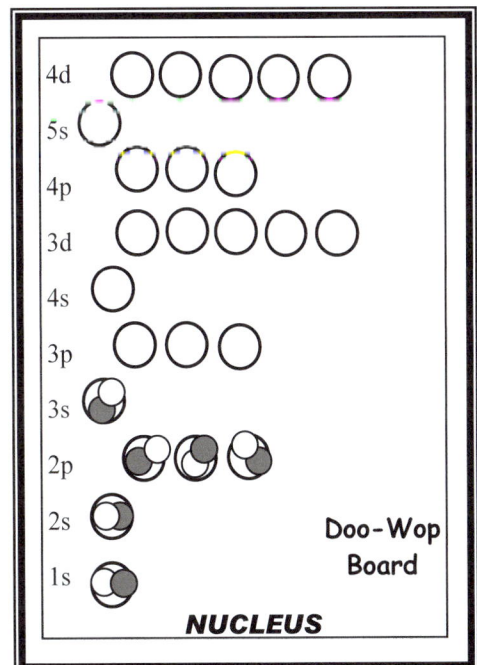

12
Mg
Magnesium
24.31

Doo-Wop Board
NUCLEUS

S63

Note that the electrons in the outermost orbit would be the 3s electrons.

$$Mg: \; 1s^2 \; 2s^2 \; 2p^6 \; \boxed{3s^2}$$

The number of valence electrons in magnesium would be 2 and we would write the electron dot notation as follows:

$$Mg:$$

Let's practice with one more example.

Example 3: Write the electron dot notation for <u>argon</u>.

Begin by writing the electronic configuration notation for argon here. Refer to your periodic table of elements and your Doo-Wop Board to help you write the electronic configuration, if necessary.

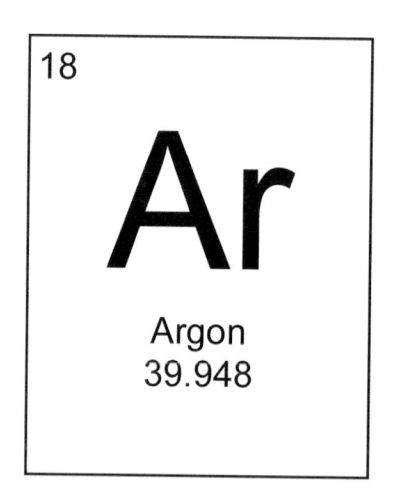

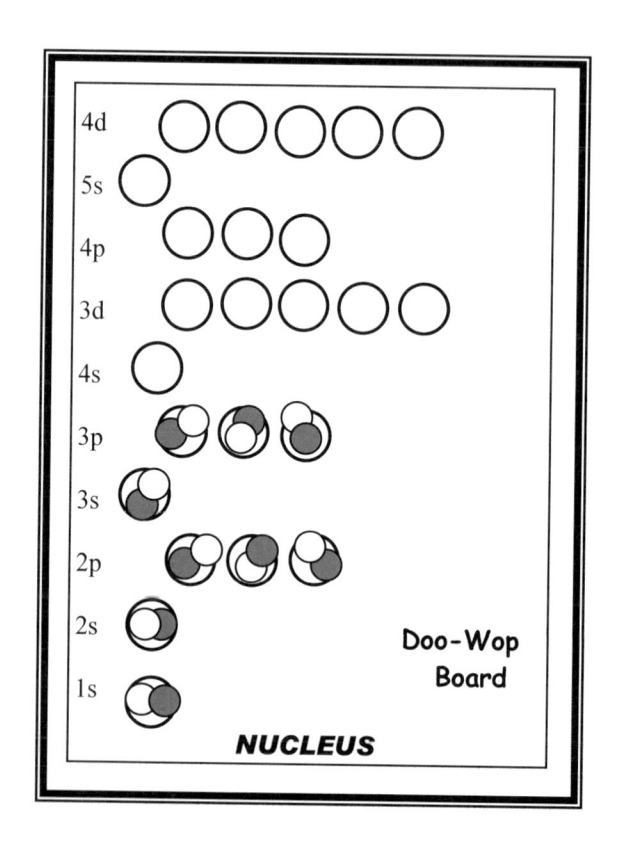

Write the electron configuration notation for argon here:

Go back now and circle the outermost orbits. How many valence electrons does argon have?

If you answered eight, you are correct! If not, count again. Note there are 2 electrons in the 3s orbit plus 6 in the filled 3p orbits for a total of eight valence electrons.

Therefore, the dot notation for argon will be (write it here!):

(You should have written 8 dots around the symbol Ar.)

To practice writing more electron dot notations, complete the practice pages which follow, play the electron dot notation version of the Teamwork Game and Level 4 of Doo-Wop Mania.

Friendly Chemistry

Name_____ Date_____

Friendly Chemistry

Electron Dot Notation Practice 1

Write the electronic configuration notation (ECN) and dot notation for each element listed below.

Element	Electronic Configuration Notation (ECN)	Electron Dot Notation
1. Carbon		
2. Boron		
3. Chlorine		
4. Neon		
5. Fluorine		
6. Beryllium		
7. Sodium		
8. Aluminum		
9. Silicon		
10. Potassium		
11. Magnesium		
12. Chromium		
13. Titanium		
14. Lithium		
15. Manganese		
16. Vanadium		
17. Phosphorus		
18. Arsenic		
19. Scandium		
20. Hydrogen		

Name_____ Date_____

Friendly Chemistry

Electron Dot Notation Practice 2

Write the electron dot notation for each element listed below.

Element	Electron Dot Notation
1. Neon	
2. Aluminum	
3. Chlorine	
4. Phosphorus	
5. Manganese	
6. Oxygen	
7. Sodium	
8. Gallium	
9. Iodine	
10. Potassium	
11. Scandium	
12. Cobalt	
13. Zinc	
14. Copper	
15. Molybdenum	
16. Calcium	
17. Silicon	
18. Sulfur	
19. Argon	
20. Helium	

Lesson 8: Relating Electron Arrangement to Reactivity

In Lessons 5, 6 and 7 we discussed how to take the information from your Doo-Wop Board and, using the quantum mechanics "code," describe the arrangement of electrons in an atom with **orbital, electron configuration** and **electron dot notations**. In this lesson we will take those descriptive notations and relate them to the reactivity, or lack of reactivity, of various elements in the periodic table. In addition, we will look at other properties of elements and begin to describe why the elements are arranged as they are in the periodic table.

Let's begin by examining the electron dot notations of a particular set of elements. Using the information presented in the last lessons, write the electron dot notations for the following elements. Remember to write the electron configuration notation first and then represent the valence electrons (those electrons in the outermost energy level or orbit) as dots.

Hydrogen
Electron configuration notation:

Electron dot notation:

Lithium
Electron configuration notation:

Electron dot notation:

Sodium
Electron configuration notation:

Electron dot notation:

Potassium
Electron configuration notation:

Electron dot notation:

Rubidium
Electron configuration notation:

Electron dot notation:

Did you notice anything in particular after completing the electron dot notations for these five elements? Yes, each of these elements has **one** valence electron. Now, find these elements on your periodic table of elements. Did you find something else that these elements hold in common? Yes, these elements are found in the farthest left column of the periodic table of elements. Columns of elements on the periodic table are known as **families** or **groups**.

Families of elements are found in columns on the periodic table.

Friendly Chemistry

The term family is more descriptive in that, possibly like your family, its members generally resemble each other. In other words, if you have biological brothers or sisters, you can usually see similarities in hair color, facial features or body build among you and your brothers or sisters.

In an analogous way, the elements which are within the same column on the periodic table of elements, have similar features or properties and are known as families of elements. Just as your family has a last name, the families on the periodic table of elements have names. The family of elements you just examined is known as the **sodium family** (that is, hydrogen, lithium, sodium, potassium, rubidium, cesium and francium are members of the sodium family).

The Sodium Family members here are shaded in gray.

Let's look at another family of elements. Write the electron dot notations for the following elements.

Neon
Electron configuration notation:

Electron dot notation:

Argon
Electron configuration notation:

Electron dot notation:

Krypton
Electron configuration notation:

Electron dot notation:

Again, can you see what feature is common to this set of elements? Yes, each of the elements in this family of elements has eight valence electrons. Look at your periodic table to find these elements. Do you see them on the far right side of your periodic table? This family of elements is known as the **noble gas family** or **inert gas family**.

Note that helium is also a member of this family of elements. If you write the electron dot notation for helium, you will find that helium has only two valence electrons and not eight like the other members in the family. Later in this chapter we will discuss why helium is included in the noble gas family.

Another family to be "introduced" at this point consists of the following elements. Write the electron dot notation for these elements and then find them on your periodic table.

Fluorine

Electron configuration notation:

Electron dot notation:

Chlorine

Electron configuration notation:

Electron dot notation:

Bromine

Electron configuration notation:

Electron dot notation:

Friendly Chemistry

How many valence electrons do each of these elements have? If you said seven, you are correct! This family of elements can be found just to the left of the noble gas family. The name of this family of elements is the **halogens**. Notice that each family member's name ends in -ine (fluor_ine_, chlor_ine_, brom_ine_, iod_ine_ and astat_ine_).

The Halogen Family members are shaded in gray.

									2 He Helium 4.0026
				5 B Boron 10.811	6 C Carbon 12.0115	7 N Nitrogen 14.0067	8 O Oxygen 15.994	9 F Fluorine 18.994	10 Ne Neon 20.18
				13 Al Aluminum 26.9815	14 Si Silicon 28.086	15 P Phosphorus 30.974	16 S Sulfur 32.06	17 Cl Chlorine 35.453	18 Ar Argon 39.948
27 Co Cobalt 58.933	28 Ni Nickel 58.71	29 Cu Copper 63.546	30 Zn Zinc 65.37	31 Ga Gallium 69.72	32 Ge Germanium 72.59	33 As Arsenic 74.9216	34 Se Selenium 78.96	35 Br Bromine 79.909	36 Kr Krypton 83.80
45 Rh Rhodium 102.91	46 Pd Palladium 106.4	47 Ag Silver 107.868	48 Cd Cadmium 112.40	49 In Indium 114.82	50 Sn Tin 118.69	51 Sb Antimony 121.75	52 Te Tellurium 127.60	53 I Iodine 126.904	54 Xe Xenon 131.30
77 Ir Iridium 192.22	78 Pt Platinum 195.09	79 Au Gold 196.97	80 Hg Mercury 200.59	81 Tl Thallium 204.37	82 Pb Lead 207.2	83 Bi Bismuth 208.98	84 Po Polonium (210)	85 At Astatine (210)	86 Rn Radon (222)
109 Une									

Let's continue by examining another family of elements. Write the electron dot notations for each of the elements listed below and then find them on your periodic table.

Oxygen
Electron configuration notation:

Electron dot notation

S75

Sulfur

Electron configuration notation:

Electron dot notation:

Selenium

Electron configuration notation:

Electron dot notation:

    How many valence electrons do each of these elements have? If you said six, you're right! Look at the periodic table below to see where this family of elements "resides." The name of this family of elements is the **oxygen family**. The members of the oxygen family share properties unique to that family.

Friendly Chemistry

One additional family that we will examine consists of the following elements. Write the electron dot notation for each of the elements below and then find them on your periodic table of elements.

Beryllium
Electron configuration notation:

Electron dot notation:

Magnesium
Electron configuration notation:

Electron dot notation:

Calcium
Electron configuration notation:

Electron dot notation:

Strontium
Electron configuration notation:

Electron dot notation:

Did you find this set of elements "next door" to the sodium family? This set of elements, with two valence electrons each, is known as the **calcium family**.

The Calcium Family members are shaded in gray.

S77

Review the periodic table of elements below to see the five families we have discussed so far. Take a moment to label these element families on your own periodic table of elements. Then, practice learning the family names by playing Element Family Demo Derby!

Friendly Chemistry

Name_____ Date_____

Friendly Chemistry

Element Family Practice 1

Write the dot notation and then tell the family to which the elements listed below belong.

Element Name	Electron Dot Notation	Element Family
1. Sodium		
2. Oxygen		
3. Helium		
4. Potassium		
5. Fluorine		
6. Calcium		
7. Beryllium		
8. Hydrogen		
9. Argon		
10. Sulfur		
11. Strontium		
12. Chlorine		
13. Magnesium		
14. Iodine		
15. Lithium		
16. Astatine		
17. Barium		
18. Krypton		
19. Xenon		
20. Rubidium		

S79

Name_____ Date_____

Friendly Chemistry

Element Family Practice 2

Write the dot notation and then tell the family to which the elements listed below belong.

Element Name	Electron Dot Notation	Element Family
1. Silicon		
2. Helium		
3. Sulfur		
4. Argon		
5. Calcium		
6. Beryllium		
7. Fluorine		
8. Neon		
9. Krypton		
10. Bromine		
11. Hydrogen		
12. Potassium		
13. Lithium		
14. Oxygen		
15. Selenium		
16. Strontium		
17. Iodine		
18. Xenon		
19. Cesium		
20. Enough practice for one day!	This one's a free one!	We'll just call it an early Christmas present!

Lesson 9: Another Trend of the Periodic Table: Ionization Energy

In Lesson 8, we found that elements were grouped into families on the periodic table according to the number of valence electrons they had. We found that members of the sodium family all had one valence electron, members of the calcium family had two valence electrons and members of the oxygen family and halogen family had six and seven valence electrons, respectively. Additionally, we found that the noble gas family members had eight valence electrons (except for helium). In this lesson, we will continue to explore reasons why elements are grouped into families on the periodic table. This time we will look at a property called ionization energy.

We have said that the reactivity of an element is based upon its arrangement of electrons. When we discussed electron dot notation, we learned that the electrons in the outermost orbits are called valence electrons. We also said the valence electrons were available for creating bonds with other atoms.

Friendly Chemistry

In some cases, these valence electrons actually move to another atom to create the bond or are shared between two (or more) atoms to create the bond. The ease with which these electrons move from atom to atom is directly related to the reactivity of any element. In other words, if the valence electrons can easily be moved or transferred to another atom, that element is said to be quite reactive compared to an element in which the valence electrons cannot be easily moved or transferred to another atom.

Moving or transferring electrons requires energy. The energy required to *remove* electrons is given the name **ionization energy**. Ionization refers to the creation of an ion. An **ion** is what an atom is called when it either loses or gains electrons in the process of creating a bond with another atom. Therefore, ionization energy is the energy required to remove an electron from an atom.

Energy required to *add* electrons to atoms is called **electron affinity**. So, in addition to the degree of reactivity, the amount of ionization energy and the amount of electron affinity are two other properties by which elements are grouped into families. To better visualize this concept, let's create a graph which illustrates the varying amounts of ionization energy among the elements in the periodic table.

Below is a chart which lists the relative amounts of ionization energy required to remove one electron from an atom of that particular element. On the next page is the beginning of a line graph to which the information on the chart will be transferred. Take some time and complete the graph. Note that you go down each column on the chart.

Ionization Energy Values for elements 1-49.

Element	Ionization Energy	Element	Ionization Energy	Element	Ionization Energy
Hydrogen	314	Argon	363	Selenium	225
Helium	567	Potassium	100	Bromine	272
Lithium	124	Calcium	141	Krypton	323
Beryllium	215	Scandium	151	Rubidium	96
Boron	191	Titanium	157	Strontium	131
Carbon	260	Vanadium	155	Yttrium	147
Nitrogen	335	Chromium	156	Zirconium	158
Oxygen	314	Manganese	171	Niobium	159
Fluorine	402	Iron	181	Molybdenum	164
Neon	497	Cobalt	181	Technetium	168
Sodium	119	Nickel	176	Ruthenium	170
Magnesium	176	Copper	178	Rhenium	172
Aluminum	138	Zinc	217	Palladium	192
Silicon	188	Gallium	138	Silver	175
Phosphorus	242	Germanium	182	Cadmium	207
Sulfur	239	Arsenic	226	Indium	133
Chlorine	299				

Friendly Chemistry

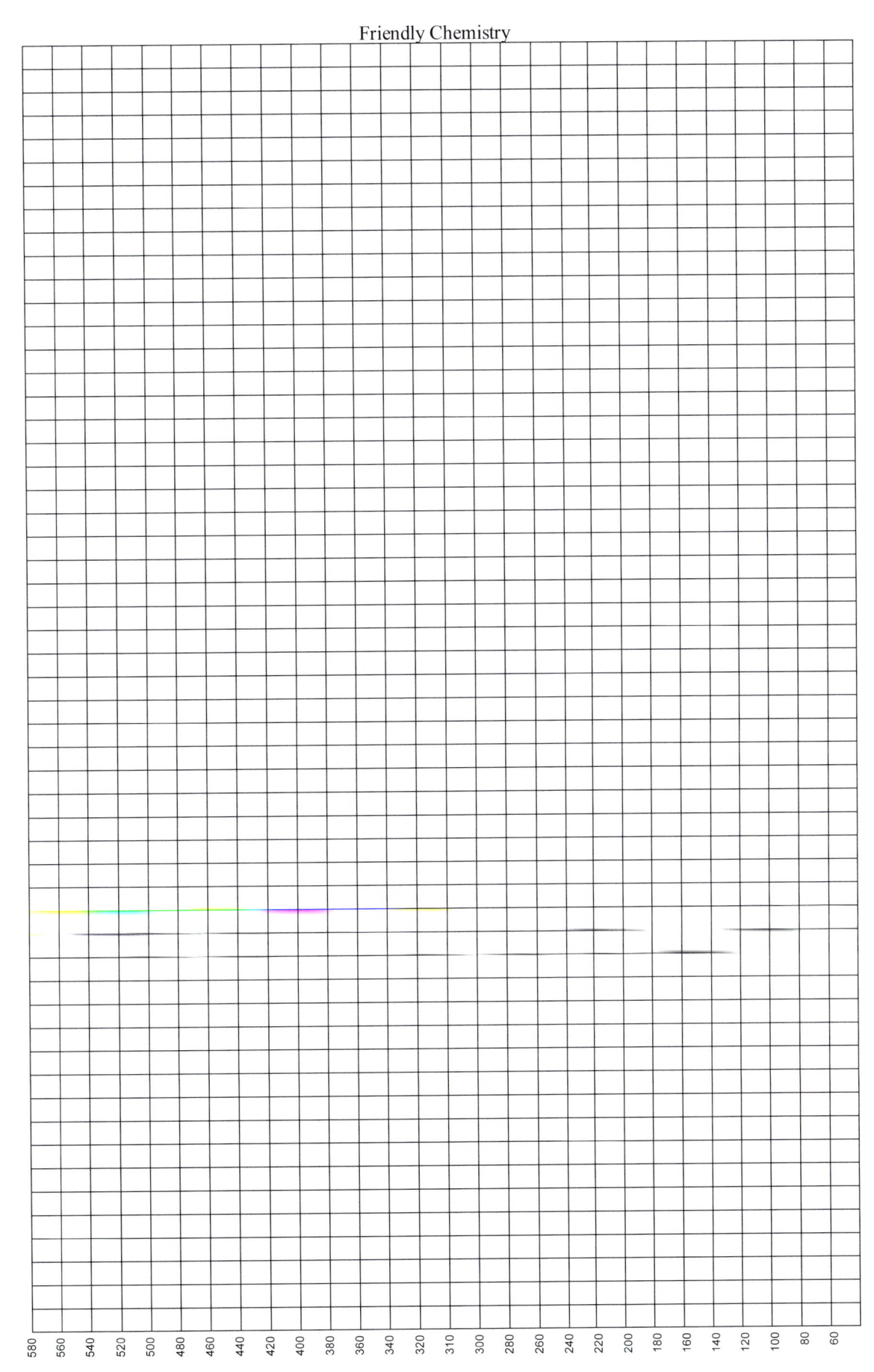

H He Li Be

580 560 540 520 500 480 460 440 420 400 380 360 340 320 310 300 280 260 240 220 200 180 160 140 120 100 80 60

S83

After graphing the ionization energy values of the first 49 elements, go back to your graph and find the elements that correspond to the <u>peaks</u> on your graph. Write those element symbols and names here:

Symbol	Element Name

Now, look back at your periodic table of elements to find these elements. Are they in a particular family of elements with which you are familiar? To which family do they belong? _____ Yes, they are members of the noble gas family!

Based upon the ionization energy values of these elements (including xenon and radon), it can be stated that it is extremely difficult to remove an electron from an atom of a noble gas. In other words, members of the noble gas family are considered very non-reactive. The noble gases are also known as inert gases (inert meaning stable or non-reactive).

Take the noble gas helium, for example. Helium, as you are probably aware, has a density which is less than the density of the gases which make up the air we breathe which is a mixture primarily of oxygen and nitrogen. This low density explains why helium-filled balloons will rise on a string. Because helium is so non-reactive, helium-filled balloons are relatively safe for children. Helium is also used to fill airships, which are also known as blimps and dirigibles. However, this was not always the case! Let's examine the reactivity of another family of elements to find out why helium has replaced another element that was first used in airships.

Look back at your graph showing ionization energy of the first 49 elements. Find the lowest points (valleys) on the graph which correspond to the elements with the lowest ionization energies. Write those symbols and element names here.

Symbol	Element Name
H	Hydrogen

While hydrogen may not appear to be a "valley" on your graph, it should be included in this family of elements.

Find this group of elements on your periodic table. Which family of elements do they belong to? _____ As a family, these elements (hydrogen, lithium, sodium, potassium, rubidium, cesium and francium) share the property of having the *lowest* ionization energy. This means that it takes a relatively low amount of energy to remove one electron from the outer electron orbits of atoms of these elements. In other words, these elements are by far the **most reactive elements known to man!** And quite a "rowdy" bunch of elements they are!

Lets return to the history of airships. Accounts of early air travel via airship or blimp tell us that due to its low density, hydrogen gas was used to raise airships off the ground. However, hydrogen, as we have just learned, is a member of a highly reactive family of elements. Because hydrogen was used to fill the famous German-built *Hindenburg* airship, the airship was destroyed by fire in 1937 as it was landing at Lakehurst, New Jersey. This rigid airship had a length of 245 m (804 ft) and a gas capacity of 190,006,030 liters (6,710,000 cu ft). One-third of the *Hindenburg's* passengers and crew were killed in the accident. Since that time, stable helium has replaced reactive hydrogen to get blimps and dirigibles off the ground.

Other members of the sodium family are also known for their high levels of reactivity. Sodium and potassium in their pure forms are so reactive that they must be stored "under" a petroleum product such as diesel or kerosene to prevent exposure to water vapor in the air. If sodium or potassium contact water, a violent reaction results producing strong basic compounds and hydrogen gas (the member of this rowdy bunch that we have already discussed).

To summarize this lesson, you learned that valence electrons are sometimes moved from atoms in order to create bonds between those atoms. The energy it takes to do so is called ionization energy. You then created a graph of the ionization energy values and found that the families of elements shared similar ionization energy values. We looked at two families in particular and found that the noble gas family members had very high ionization energy values, meaning it takes "tons" of energy to remove an electron, therefore making them very non-reactive elements.

On the other hand, we found that the sodium family members had very low ionization energy values making them extremely reactive. The remaining elements on the periodic table have varying degrees of ionization energy, hence they have varying degrees of reactivity associated with them.

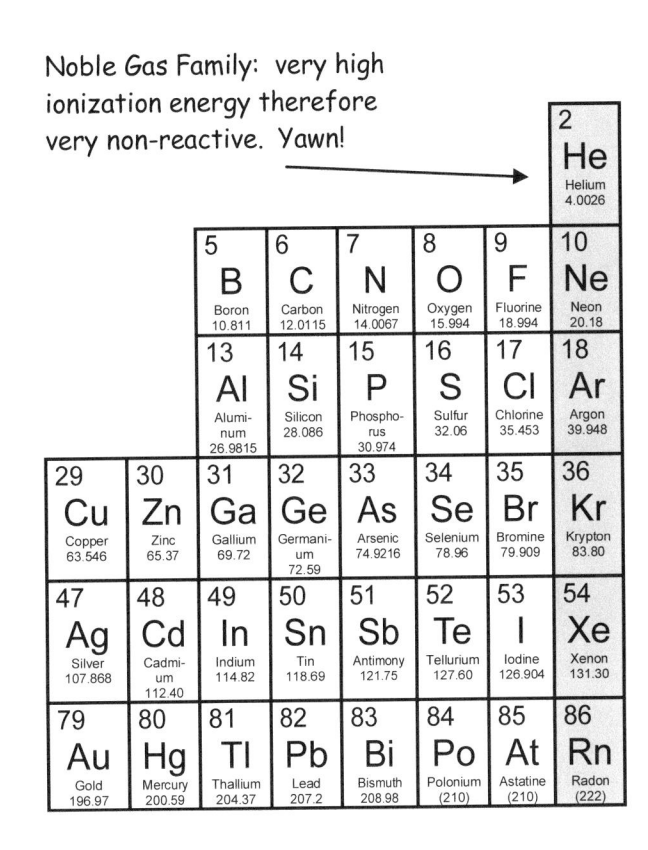

Friendly Chemistry

Sodium Family: very low ionization energy therefore very high reactivity! Watch out!!

Noble Gas Family: very high ionization energy therefore very non-reactive. Yawn!

1 H Hydrogen 1.0080																	2 He Helium 4.0026
3 Li Lithium 6.94	4 Be Beryllium 9.012											5 B Boron 10.811	6 C Carbon 12.0115	7 N Nitrogen 14.0067	8 O Oxygen 15.994	9 F Fluorine 18.994	10 Ne Neon 20.18
11 Na Sodium 22.9898	12 Mg Magnesium 24.31											13 Al Aluminum 26.9815	14 Si Silicon 28.086	15 P Phosphorus 30.974	16 S Sulfur 32.06	17 Cl Chlorine 35.453	18 Ar Argon 39.948
19 K Potassium 39.102	20 Ca Calcium 40.08	21 Sc Scandium 44.96	22 Ti Titanium 47.9	23 V Vanadium 50.94	24 Cr Chromium 51.996	25 Mn Manganese 54.938						29 Cu Copper 63.546	30 Zn Zinc 65.37	31 Ga Gallium 69.72	32 Ge Germanium 72.59	33 As Arsenic 74.9216	34 Se Selenium 78.96	35 Br Bromine 79.909	36 Kr Krypton 83.80
37 Rb Rubidium 85.47	38 Sr Strontium 87.62	39 Y Yttrium 88.91	40 Zr Zirconium 91.22	41 Nb Niobium 92.91	42 Mo Molybdenum 95.94	43 Tc Technetium (99)						47 Ag Silver 107.868	48 Cd Cadmium 112.40	49 In Indium 114.82	50 Sn Tin 118.69	51 Sb Antimony 121.75	52 Te Tellurium 127.60	53 I Iodine 126.904	54 Xe Xenon 131.30
55 Cs Cesium 132.91	56 Ba Barium 137.34	71 Lu Lutetium 174.97	72 Hf Hafnium 178.49	73 Ta Tantalum 180.95	74 W Tungsten 183.85	75 Re Rhenium 186.2						79 Au Gold 196.97	80 Hg Mercury 200.59	81 Tl Thallium 204.37	82 Pb Lead 207.2	83 Bi Bismuth 208.98	84 Po Polonium (210)	85 At Astatine (210)	86 Rn Radon (222)
87 Fr Francium (rra)	88 Ra Radium (226)	103 Lr Lawrencium (259)	104 Unq	105 Unp	106 Unh	107 Uns											

NAME_____ DATE_____

FRIENDLY CHEMISTRY

Lesson 9 Practice Page
Element Reactivity and Ionization Energy

Fill in the blanks with an appropriate term. Sometimes, more than one term may be correct.

In this lesson, we began with the review of _____ electrons or those electrons found in the outermost energy levels of an atom. We discussed that members of the _____ family have one valence electron while members of the calcium family have _____ valence electrons. We also refreshed our memory that members of the oxygen family have _____ valence electrons while members of the halogen family have _____ valence electrons. We also remembered that members of the _____ family have eight valence electrons (except for _____ which has _____ valence electrons).

Next, we played a "game" where our instructor either held onto a ball tightly or hardly at all. By playing this game, we learned about _____ which is the energy required to remove an electron from an atom. When our instructor held the ball very tightly, he/she was demonstrating _____ ionization energy values. We learned that members of the _____ have really high ionization energy values. This makes them very _____. Another name for this family is the _____ gas family (where inert means non-reactive). Members of this family include: _____
_____.

When our instructor held the ball very loosely or hardly at all, he/she was demonstrating _____ ionization energy. This means that the elements with _____ ionization energy are very willing to _____ with other elements. They are said to be highly _____. The family of elements with very low ionization energy values is the _____ family. Members of this family include:
_____. The family member _____, for example is so very reactive that it must be stored where it cannot come into contact with _____. Exposure to water will cause a violent reaction to occur! Another member of this family was once used to make airships float due to its low density. This family member is _____. But due to its high level of _____ it is too dangerous to use.

In summary, we learned that _____ was another way elements are placed into _____ on the periodic table.

Lesson 10: A Final Trend of the Periodic Table: Atom Size

How large are the atoms of elements found on the periodic table? While we understand that individual atoms are too small for us to see, chemists have devised ways to measure how large they are in relationship to each other. In this lesson, we will examine how the atom sizes vary as you move *across* the table through elements which make up a **period** and also how size varies as you move *down* through members of a **family**. The table on the next page tells the approximate size of atoms of each element measured from the nucleus to the outer edge in units called angstroms. This measurement is known as the atomic radius. Look at the diagram below to help you understand the measurement of atomic radius. Use the information in the table to complete the graph on the following page like you did when you graphed ionization energy values in Lesson 9.

The atomic radius is measured from the nucleus to the outermost layer of electrons.

Element	Atomic Radius	Element	Atomic Radius	Element	Atomic Radius
Hydrogen	0.32	Potassium	2.03	Rubidium	2.16
Helium	0.31	Calcium	1.74	Strontium	1.91
Lithium	1.23	Scandium	1.44	Yttrium	1.62
Beryllium	0.89	Titanium	1.32	Zirconium	1.45
Boron	0.82	Vanadium	1.22	Niobium	1.34
Carbon	0.77	Chromium	1.18	Molybdenum	1.30
Nitrogen	0.74	Manganese	1.17	Technetium	1.27
Oxygen	0.70	Iron	1.17	Ruthenium	1.25
Fluorine	0.68	Cobalt	1.16	Rhenium	1.25
Neon	0.67	Nickel	1.15	Palladium	1.28
Sodium	1.54	Copper	1.17	Silver	1.34
Magnesium	1.36	Zinc	1.25	Cadmium	1.48
Aluminum	1.18	Gallium	1.26	Indium	1.44
Silicon	1.11	Germanium	1.22	Tin	1.72
Phosphorus	1.06	Arsenic	1.20	Antimony	1.53
Sulfur	1.11	Selenium	1.17	Tellurium	1.42
Chlorine	0.99	Bromine	1.14	Iodine	1.32
Argon	0.98	Krypton	1.12	Xenon	1.24

To see how the size of the elements varies as you move across a series of elements on the periodic table of elements, look at your graph and find the series of elements which begins with potassium and ends with krypton. Note that a series of elements is a row of elements going from left to right across the table. Note that potassium is at a peak and as you move across through the series the radius tends to decrease until you get to the smallest atom in the series which is krypton

It is theorized that the atoms decrease in size as you move across through a series because of the increased attractive forces between the nucleus and the electron cloud. This is thought to be due to the increased number of positively charged protons in the nucleus and negatively charged electrons in orbit. Like opposite poles on a magnet which attract each other, opposite charges in an atom attract each other. As you increase the number of oppositely charged particles (protons and electrons) the force between them tends to increase which results in the electron cloud being more compact.

Friendly Chemistry

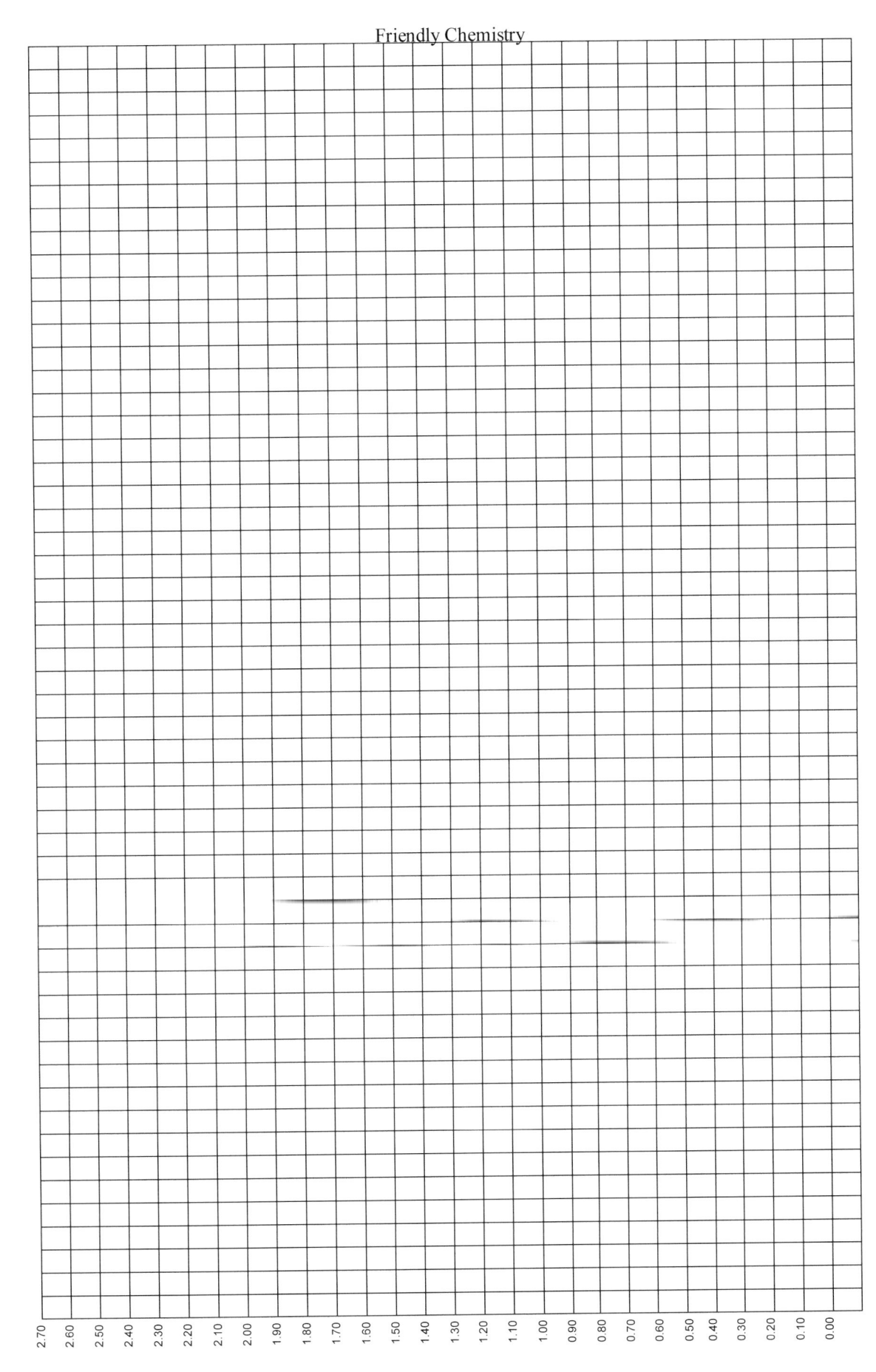

H He Li Be

2.70 2.60 2.50 2.40 2.30 2.20 2.10 2.00 1.90 1.80 1.70 1.60 1.50 1.40 1.30 1.20 1.10 1.00 0.90 0.80 0.70 0.60 0.50 0.40 0.30 0.20 0.10 0.00

S91

The overall trend is as you move *across* the table through a series of elements, atomic size tends to decrease. Let's examine how atomic size varies as you move *down* through a family of elements. Recall that as you move down through a family of elements, an additional energy level is added with each downward step you take. Look at the sodium family, for example.

H: 1s

Li: 1s 2s

Na: 1s 2s 2p 3s

K: 1s 2s 2p 3s 3p 4s

Rb: 1s 2s 2p 3s 3p 4s 3d 4p 5s

Cs: 1s 2s 2p 3s 3p 4s 3d 4p 5s 4d 5p 6s

Note in the electron configuration notations here, that with each step DOWN through the sodium family, one additional energy level is added.

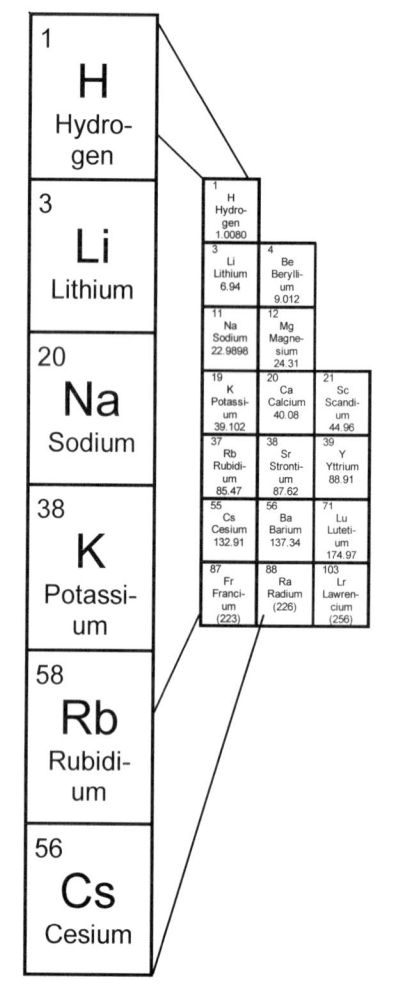

Now, examine the atomic size of each member of the sodium family by finding them on your graph from the previous page. Note that as you move down through the sodium family, with each step, the radius measurement grows. Theorists believe that with each addition of energy level, the radius increases. In other words, as each energy level is added, the size of the atom increases. Therefore, as you move down through members within the same family, the atomic radius increases.

Size **increases** as you move down through a family of elements.

Size **decreases** as you move across through a series or period of elements.

1 H Hydrogen 1.0080																	2 He Helium 4.0026
3 Li Lithium 6.94	4 Be Beryllium 9.012										5 B Boron 10.811	6 C Carbon 12.0115	7 N Nitrogen 14.0067	8 O Oxygen 15.994	9 F Fluorine 18.994	10 Ne Neon 20.18
11 Na Sodium 22.9898	12 Mg Magnesium 24.31										13 Al Aluminum 26.9815	14 Si Silicon 28.086	15 P Phosphorus 30.974	16 S Sulfur 32.06	17 Cl Chlorine 35.453	18 Ar Argon 39.948
19 K Potassium 39.102	20 Ca Calcium 40.08	21 Sc Scandium 44.96	22 Ti Titanium 47.9	23 V Vanadium 50.94	24 Cr Chromium 51.996	25 Mn Manganese 54.938	26 Fe Iron 55.847	27 Co Cobalt 58.933	28 Ni Nickel 58.71	29 Cu Copper 63.546	30 Zn Zinc 65.37	31 Ga Gallium 69.72	32 Ge Germanium 72.59	33 As Arsenic 74.9216	34 Se Selenium 78.96	35 Br Bromine 79.909	36 Kr Krypton 83.80
37 Rb Rubidium 85.47	38 Sr Strontium 87.62	39 Y Yttrium 88.91	40 Zr Zirconium 91.22	41 Nb Niobium 92.91	42 Mo Molybdenum 95.94	43 Tc Technetium (99)	44 Ru Ruthenium 101.07	45 Rh Rhodium 102.91	46 Pd Palladium 106.4	47 Ag Silver 107.868	48 Cd Cadmium 112.40	49 In Indium 114.82	50 Sn Tin 118.69	51 Sb Antimony 121.75	52 Te Tellurium 127.60	53 I Iodine 126.904	54 Xe Xenon 131.30
55 Cs Cesium 132.91	56 Ba Barium 137.34	71 Lu Lutetium 174.97	72 Hf Hafnium 178.49	73 Ta Tantalum 180.95	74 W Tungsten 183.85	75 Re Rhenium 186.2	76 Os Osmium 190.2	77 Ir Iridium 192.22	78 Pt Platinum 195.09	79 Au Gold 196.97	80 Hg Mercury 200.59	81 Tl Thallium 204.37	82 Pb Lead 207.2	83 Bi Bismuth 208.98	84 Po Polonium (210)	85 At Astatine (210)	86 Rn Radon (222)
87 Fr Francium (223)	88 Ra Radium (226)	103 Lr Lawrencium (256)	104 Unq	105 Unp	106 Unh	107 Uns	108 Uno	109 Une									

To get a 3-D view of how size varies, make an atomic radius cake. Your teacher will have the instructions. Following that activity, reinforce what you've learned by completing the Lesson 10 Practice Page.

Periodic Table Of the Elements

Friendly Chemistry

Legend:
- Atomic Number → 1
- Element Symbol → H
- Element Name → Hydrogen
- Atomic Mass → 1.0080

1 H Hydrogen 1.0080																	2 He Helium 4.0026
3 Li Lithium 6.94	4 Be Beryllium 9.012											5 B Boron 10.811	6 C Carbon 12.0115	7 N Nitrogen 14.0067	8 O Oxygen 15.994	9 F Fluorine 18.994	10 Ne Neon 20.18
11 Na Sodium 22.9898	12 Mg Magnesium 24.31											13 Al Aluminum 26.9815	14 Si Silicon 28.086	15 P Phosphorus 30.974	16 S Sulfur 32.06	17 Cl Chlorine 35.453	18 Ar Argon 39.948
19 K Potassium 39.102	20 Ca Calcium 40.08	21 Sc Scandium 44.96	22 Ti Titanium 47.9	23 V Vanadium 50.94	24 Cr Chromium 51.996	25 Mn Manganese 54.938	26 Fe Iron 55.847	27 Co Cobalt 58.933	28 Ni Nickel 58.71	29 Cu Copper 63.546	30 Zn Zinc 65.37	31 Ga Gallium 69.72	32 Ge Germanium 72.59	33 As Arsenic 74.9216	34 Se Selenium 78.96	35 Br Bromine 79.909	36 Kr Krypton 83.80
37 Rb Rubidium 85.47	38 Sr Strontium 87.62	39 Y Yttrium 88.91	40 Zr Zirconium 91.22	41 Nb Niobium 92.91	42 Mo Molybdenum 95.94	43 Tc Technetium (99)	44 Ru Ruthenium 101.07	45 Rh Rhodium 102.91	46 Pd Palladium 106.4	47 Ag Silver 107.868	48 Cd Cadmium 112.40	49 In Indium 114.82	50 Sn Tin 118.69	51 Sb Antimony 121.75	52 Te Tellurium 127.60	53 I Iodine 126.904	54 Xe Xenon 131.30
55 Cs Cesium 132.91	56 Ba Barium 137.34	71 Lu Lutetium 174.97	72 Hf Hafnium 178.49	73 Ta Tantalum 180.95	74 W Tungsten 183.85	75 Re Rhenium 186.2	76 Os Osmium 190.2	77 Ir Iridium 192.22	78 Pt Platinum 195.09	79 Au Gold 196.97	80 Hg Mercury 200.59	81 Tl Thallium 204.37	82 Pb Lead 207.2	83 Bi Bismuth 208.98	84 Po Polonium (210)	85 At Astatine (210)	86 Rn Radon (222)
87 Fr Francium (223)	88 Ra Radium (226)	103 Lr Lawrencium (256)	104 Unq	105 Unp	106 Unh	107 Uns	108 Uno	109 Une									

S94

NAME_____ DATE_____
FRIENDLY CHEMISTRY

Lesson 10 Practice Page - Atomic Size

Listed below are pairs of elements. By looking at your periodic table of elements, determine first if they are members of the same <u>family</u> or are found in the same <u>series</u> on the periodic table. Based upon that information, tell which of the two elements that has the greater atomic radius.

Element A	Element B	Which element has the larger atomic radius? Write it's symbol here. (The first one has been done for you!)
1. Hydrogen	Sodium	Na
2. Sodium	Silicon	
3. Potassium	Lithium	
4. Chromium	Cobalt	
5. Boron	Gallium	
6. Fluorine	Nitrogen	
7. Argon	Xenon	
8. Iron	Osmium	
9. Chromium	Calcium	
10. Krypton	Copper	
11. Rubidium	Francium	
12. Zinc	Nickel	
13. Sulfur	Tellurium	
14. Arsenic	Bromine	
15. Technetium	Manganese	
16. Magnesium	Chlorine	
17. Helium	Radon	
18. Potassium	Scandium	
19. Calcium	Chromium	
20. Chlorine	Aluminum	

S95

Lesson 11: Ion Formation

In the last few lessons we discussed how family members on the periodic table of elements share similar properties (such as arrangement of electrons) which, in turn, are directly related to the relative degree of reactivity exhibited by those particular elements. One measure of reactivity that we discussed was based upon the amount of energy required to lose or gain electrons. We called this energy **ionization energy** (when electrons were released). We called the process of gaining or losing electrons ionization and the resulting atom, an ion. However, we did not explain *why* an atom might desire to gain or lose electrons or *when* an atom might do so. This lesson will explain the process of ionization and answer those questions.

Let's begin by re-examining the noble gas family. Recall that this family of elements can be found on the far right side of the periodic table of elements. In the past lessons we discussed that members of the noble gas family have relatively high (*very high, in fact!*) ionization energy values which means it takes a relatively large amount of energy to ionize a noble gas. In other words, it is very unlikely that noble gas family members would ionize (which translates to the ideas that noble gas family members are "happy just the way they are!"). Noble gases are, as a family, very non-reactive. Recall from your earlier lessons that the **noble gases have their outer energy level or orbit completely filled (eight valence electrons).** This, in theory, makes the atoms of the noble gases very stable. Overall, the noble gases are very stable, non-reactive elements and their tendency to ionize is quite remote.

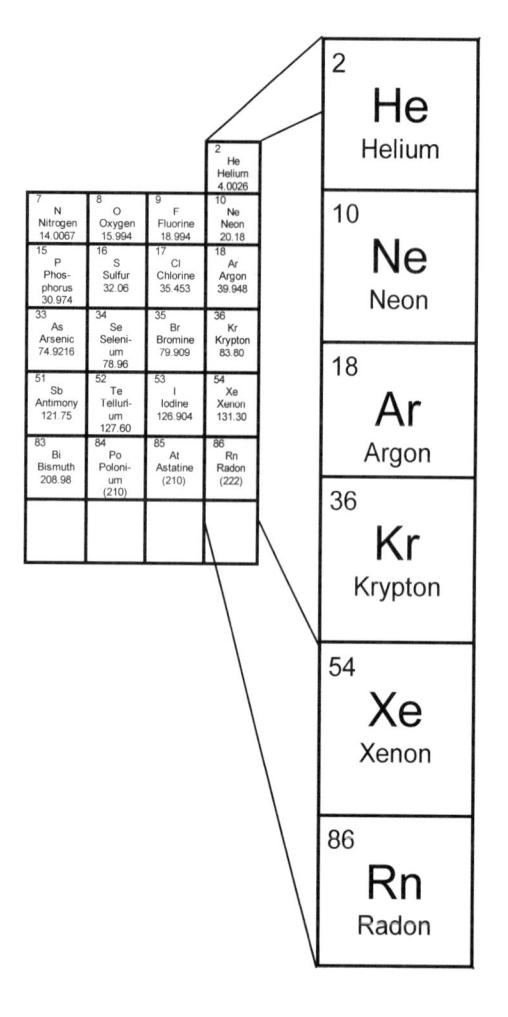

The Noble Gas or Inert Gas Family. I call these the "snob" gases—they think they're a little better than all the other elements and don't want to have anything to do (react) with any other element.

This is *not* the case for the remaining elements of the periodic table! In fact, some elements have very strong "desires" to achieve that "status" of a noble gas! To help you understand the ionization tendencies of the other families of elements, let's imagine that the families of the periodic table are like families in a fairy tale kingdom.

Imagine that the noble gas family is, indeed, the noble family of the kingdom led by Queen Helium and King Neon. Heirs to the throne include Princesses Argon and

Krypton along with Prince Xenon and Prince Radon. Together they "rule" the periodic table kingdom.

Just outside the noble gas family's castle is a family of elements that our "howlin'" to get in and take over the kingdom. This family, as we discussed in an earlier lesson, is the halogens and can be found immediately to the left of the noble gases on the periodic table.

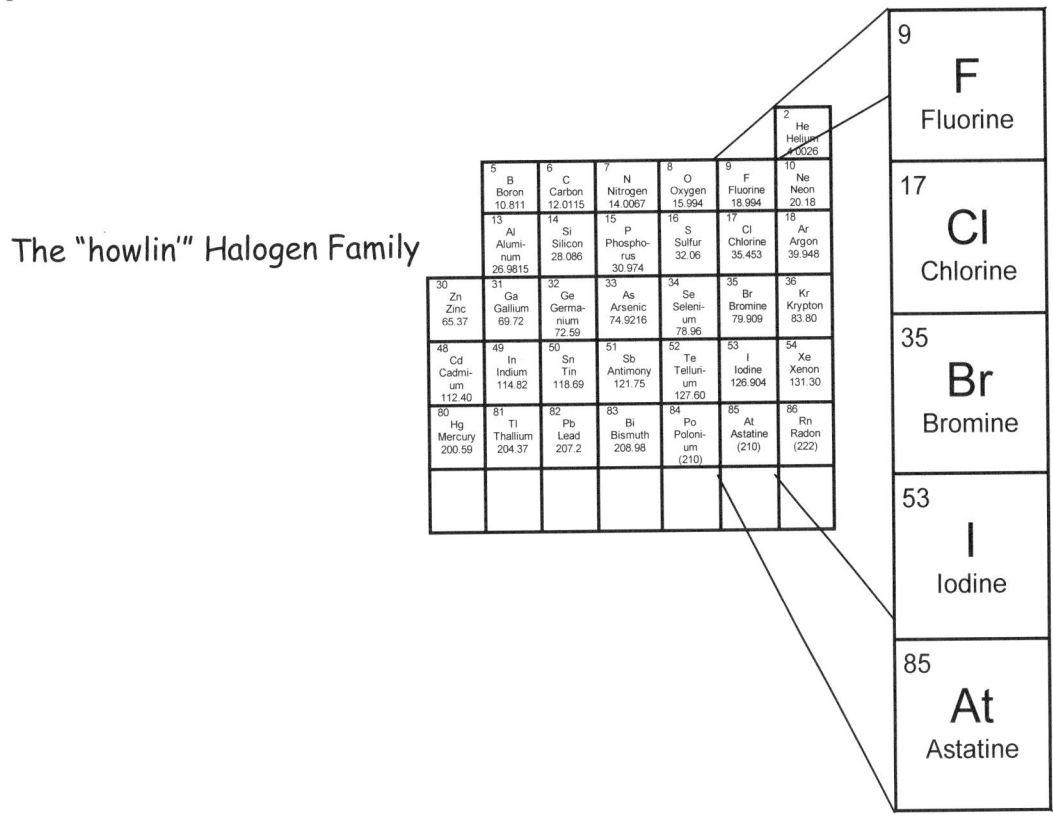

The "howlin'" Halogen Family

The elements that are members of the halogen family do not have electron arrangements like the noble gases. Sir Fluorine heads up the "howlin'" halogens. Recall that fluorine has nine total electrons. Seven of those nine electrons are in its outermost orbit (seven valence electrons). Look below to review the arrangement of electrons for fluorine.

$$F: 1s^2 \boxed{2s^2 \, 2p^5}$$

If we compare this configuration of electrons to Sir Fluorine's closest noble gas adversary (King Neon), we see that where Sir Fluorine has seven electrons in its outermost orbit (seven valence electrons), King Neon has a full complement of eight electrons (eight valence electrons). It is that **full complement of eight electrons in the outermost orbit** which gives King Neon his stability. Look below to see the ECN (electron configuration notation) for neon. Note that it indeed has eight valence electrons.

$$Ne: 1s^2 \boxed{2s^2\ 2p^6}$$

Now, Sir Fluorine and the other members of his family, Lady Chlorine and children Bromine, Iodine and Astatine, have a very strong desire to be like members of the noble family (noble gases). In order for the howlin' halogens to gain this stability, they each need to gain <u>one</u> electron. This "gained" electron will be added to the outermost energy level resulting in a total of **eight electrons in the outermost orbit** creating a much more stable atom and a much "happier" halogen. In our example of "unhappy" Sir Fluorine, he gains one electron to take on the electron configuration of King Neon. Note that while the electron configuration achieved by fluorine, as it gains one electron, is identical to neon, **it does not actually become neon—it only achieves the stability of neon.**

As we mentioned earlier, this gain (or loss) of electrons by an atoms to achieve the improved stability is known as ionization. When Sir Fluorine gains one electron to achieve the stability of King Neon, we say that fluorine ionizes. When members of the halogen family ionize, their names slightly change to indicate their "new and improved" form. Sir Fluorine becomes Sir Fluoride (fluoride) which you might recognize as being an ingredient in toothpaste. Lady Chlorine becomes Lady Chloride (chloride), and the children Bromine, Iodine and Astatine become Bromide, Iodide and Astatide. Note that the –ine ending in each of the non-ionized element names changes to –ide.

Non-ionized Form	Ionized Form
Fluorine becomes	Fluoride
Chlorine becomes	Chloride
Bromine becomes	Bromide
Iodine becomes	Iodide
Astatine becomes	Astatide

The family which resides "next door" to the howlin' halogens are members of the oxygen family headed by Sir Oxygen and Lady Sulfur. They, like the halogens, have "children" - a set of triplets, in fact—named Selenium, Tellurium and Polonium. If we examine the electron configuration of each member of the oxygen family, we see

The Oxygen Family

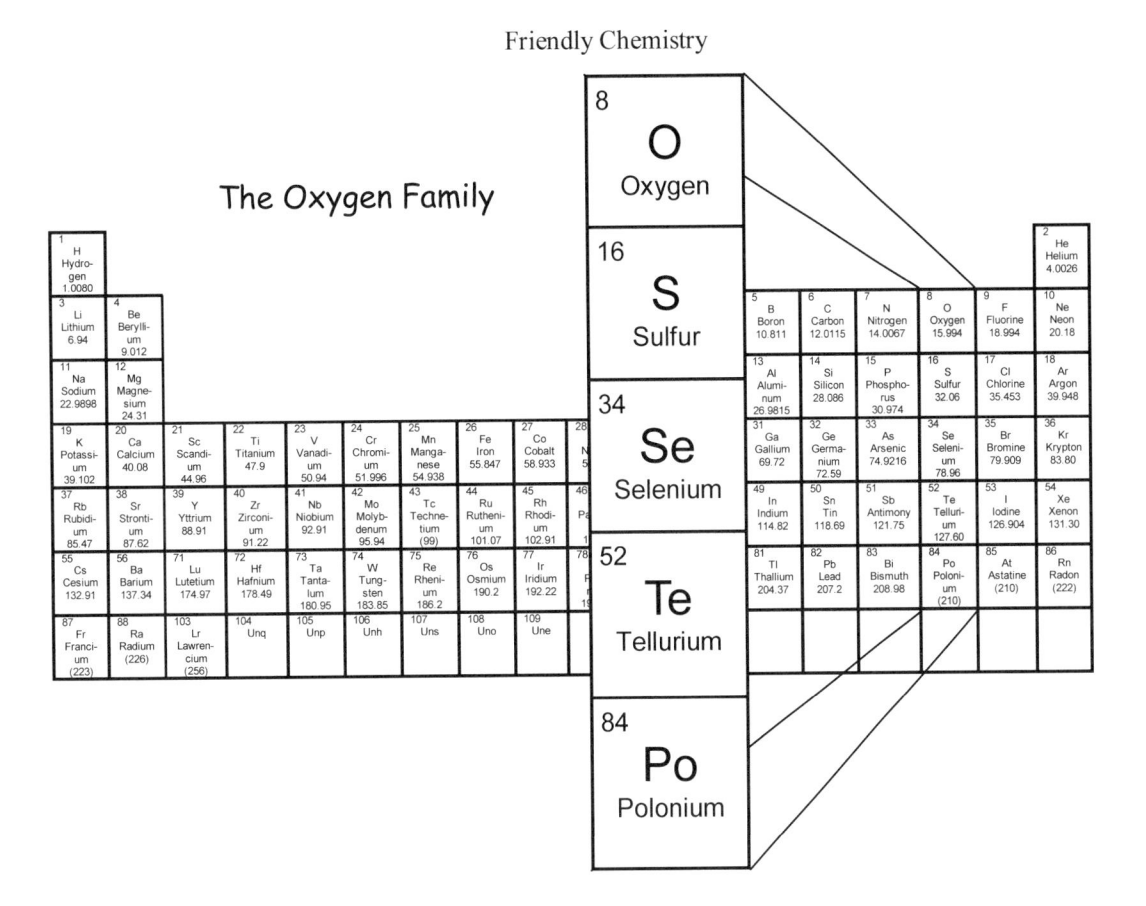

8	**O** Oxygen
16	**S** Sulfur
34	**Se** Selenium
52	**Te** Tellurium
84	**Po** Polonium

that each member has six electrons in its outermost orbit (six valence electrons). In order for members of the oxygen family to gain the stability of the noble gases, they must each gain *two* electrons.

For example, in order for oxygen to gain the stability of a noble gas, it gains two electrons to take on the configuration of neon. Remember, even though oxygen has an electron configuration similar to that of neon, it does not actually become neon. Oxygen only takes on the stability of neon as it gains those two electrons. Likewise, the other members of the oxygen family each gain two electrons to take on the configuration of their nearest noble gas family member.

As with the halogens, the members of the oxygen family also change their names to indicate that they are no ionized. Their names now also end in –ide.

Non-ionized Form	Ionized Form
Oxygen becomes	Oxide
Sulfur becomes	Sulfide
Selenium becomes	Selenide
Tellurium becomes	Telluride
Polonium becomes	Polonide

Let's now shift our attention to another family of elements which also has a very strong desire to attain the stability of the noble gases. This "rival" family of the noble gases is the sodium family. Recall that the sodium family can be found on the far left side of the periodic table of elements.

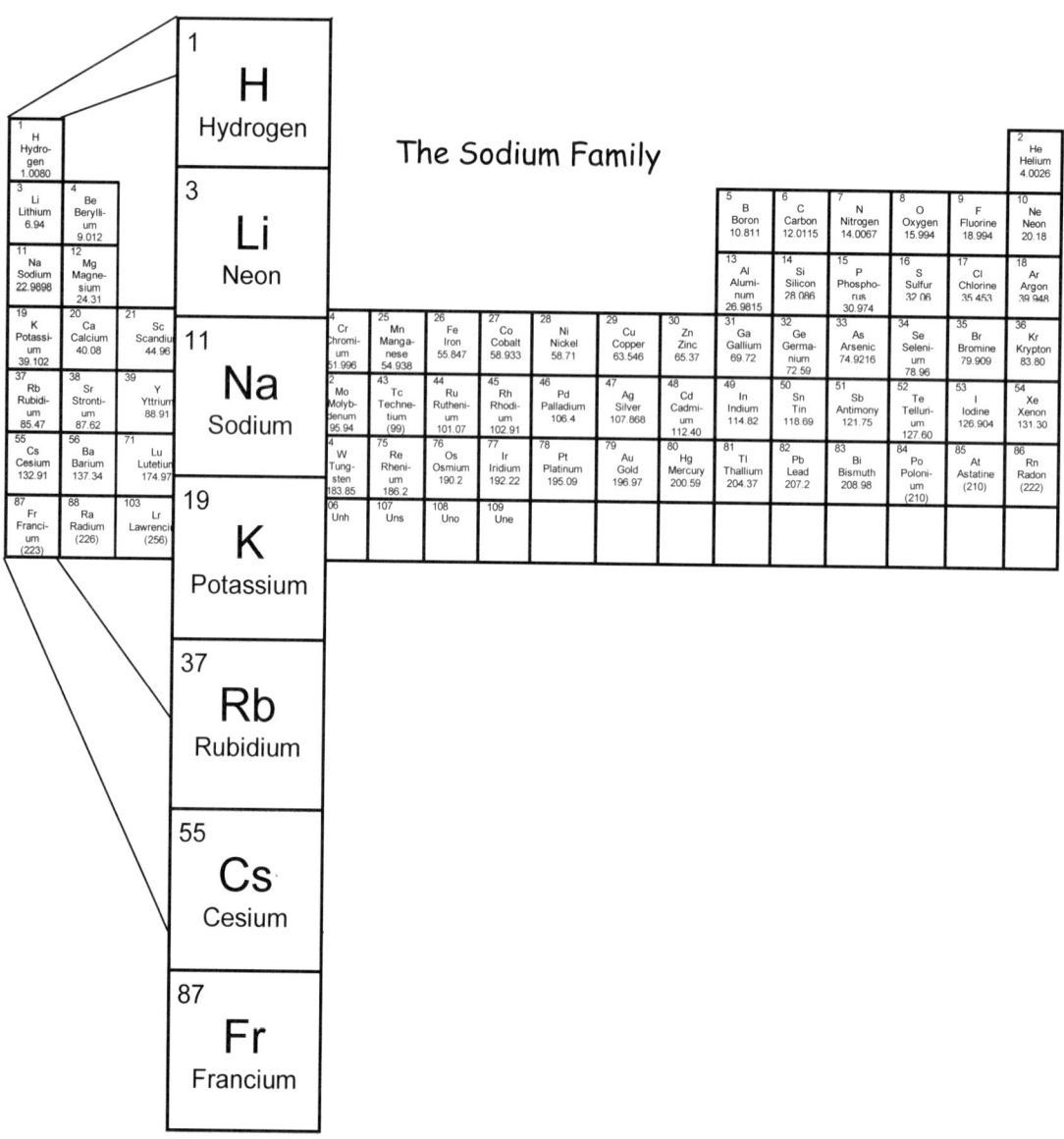

We discussed that members of the sodium family have very low ionization energy values when compared to other elements on the table. We discussed that these low values were indicative of a high degree of reactivity. By examining the electron configurations of members of the sodium family, we saw that each sodium family member has one lone electron in its outermost orbit (one valence electron).

If members of the sodium family had the opportunity to lose that one lone elec-

S102

tron in their outermost orbits, they each could attain the electron configuration of a noble gas family member. For example, lets look at lithium.

Lithium has a total of three electrons. Here is the electron configuration for lithium.

$$\text{Li:} \quad 1s^2 \boxed{2s^1}$$

Note that lithium has one valence electron in its outermost orbit. If lithium could lose that one electron it would take on the configuration of Queen Helium and be quite stable.

Another example is the element sodium. Recall that we said sodium is a highly reactive metal. If the examine its electron configuration notation, again, we see that sodium has one lone electron in its outermost orbit (one valence electron).

$$\text{Na:} \quad 1s^2 \ 2s^2 \ 2p^6 \boxed{3s^1}$$

As sodium ionizes, it loses that one lone electron to take on the electron configuration and the stability held by King Neon. The remaining sodium family members ionize in the same manner: each my losing its single valence electron. Unlike members of the halogen and oxygen families, the names of the sodium family members do *not* change as they ionize. Instead, as sodium family members ionize, their names remain in their non-ionized for and the word "ion " is added.

Non-ionized Form	Ionized Form
Hydrogen becomes the................	hydrogen ion
Lithium becomes the.................	lithium ion
Sodium becomes the.....................	sodium ion
Potassium becomes the................	potassium ion
Rubidium becomes the	rubidium ion
Cesium becomes the	cesium ion
Francium becomes the.................	francium ion

The calcium family, immediately to the right of the sodium family, also has "desires" to attain the stability of noble gas family members. Recall that members of the calcium family have two electrons in their outermost orbits (two valence electrons). As with the sodium family, the calcium family, in general, has relatively, low ionization energy values. This means that each member of the calcium family would like to lose those two valence electrons and take on the configuration of the nearest noble gas.

For example, should calcium have the opportunity to lose its two valence electrons, it could attain the configuration and stability of argon. If magnesium could lose its two valence electrons, it could attain the configuration and stability of neon. Strontium would attain the configuration and stability of krypton. The remaining calcium family members ionize in the same way—each losing its two valence electrons.

Like the sodium family members, the calcium family members' names do not change when they ionize. The word "ion" is added to their original name.

Non-ionized Form	Ionized Form
Beryllium becomes the	Beryllium ion
Magnesium becomes the	Magnesium ion
Calcium becomes the	Calcium ion
Strontium becomes the	Strontium ion
Barium becomes the	Barium ion
Radium becomes the	Radium ion

Members of the nitrogen, carbon and boron families also ionize. However, the number of electrons each family member loses is not as easy to predict as with the families we have just discussed. In fact, some of the elements (carbon, for example) do not actually *lose* electrons but rather *share* electrons with other atoms to improve the stability of both atoms involved. Similarly, the elements which make up the group known as the transition elements are quite unpredictable. Some elements (iron, for example) may lose three *or* four electrons depending on the company in which is finds itself. Mercury may lose one *or* two electrons in a likewise fashion. For these reasons, developing trends regarding the ionization of these families is not worthwhile at this point. For the sake of completing our fairytale analogy, we might call these unpredictable families the "outlaws" of the kingdom!

As a review, in this lesson we learned that members of the halogen family gain one electron as they ionize. Halogen family members' names change to end in –ide. The oxygen family members gain two electrons as they ionize and their names also change with an –ide ending. On the other hand, members of the sodium family lose one electron and members of the calcium family lose two electrons as they ionize. Unlike the halogens and oxygen families, the names of the ionized form of elements belonging to the sodium and calcium families do not change—only the term "ion" is added.

Friendly Chemistry

Name _____ Date _____

Friendly Chemistry

Lesson 11: Ion Formation Practice

Below is a table which will help you practice the ionization process of the sodium, calcium, oxygen and halogen families of elements. The first problem has been completed for you.

The element	Of the ___ family	With ___ total e's	And ___ valence e's	Gains or loses	___ e's as it ionizes	And the ion name of _____.
1. sodium	sodium	11	1	loses	1	Sodium ion
2. lithium						
3. potassium						
4. calcium						
5. barium						
6. hydrogen						
7. chlorine						
8. oxygen						
9. iodine						
10. cesium						
11. fluorine						
12. sulfur						
13. selenium						
14. polonium						
15. francium						
16. magnesium						
17. tellurium						
18. astatine						
19. strontium						
20. argon						

Lesson 12: Determining Charges on Ions

In this lesson we will discuss what happens to the overall electrostatic charge of an atom when it ionizes (gains or loses electrons). Recall from Lesson 3 that the subatomic particles within an atom carry a charge: protons are positively charged, electrons are negatively charged and neutrons are neutral.

Before an atom ionizes, is said to be electrostatically neutral (having an equal number of positively charged protons and negatively charged electrons). However, once it ionizes (gains or loses electrons) that balance becomes upset and an excess number of positive charges or negative charges results (depending if the atom had to gain or lose electrons to ionize). Let's look at an example:

Fluorine is a member of the halogen family and, as we discussed earlier, has seven valence electrons. In order for fluorine to gain a more stable electron configuration, it must gain one electron. Now, before fluorine ionizes (gains that one electron) it is electrostatically neutral (having 9 positively charged protons and 9 negatively charged electrons). By gaining one more electron, the ionized fluorine atom now has one more negative charge than the number positive charges (10 electrons vs. 9 protons). The resulting charge on the ionized fluorine is –1 and recall that its name changes to fluoride. To write the symbol and charge for the fluoride ion, we write the element symbol followed by a superscript of the resulting charge: F^{-1}.

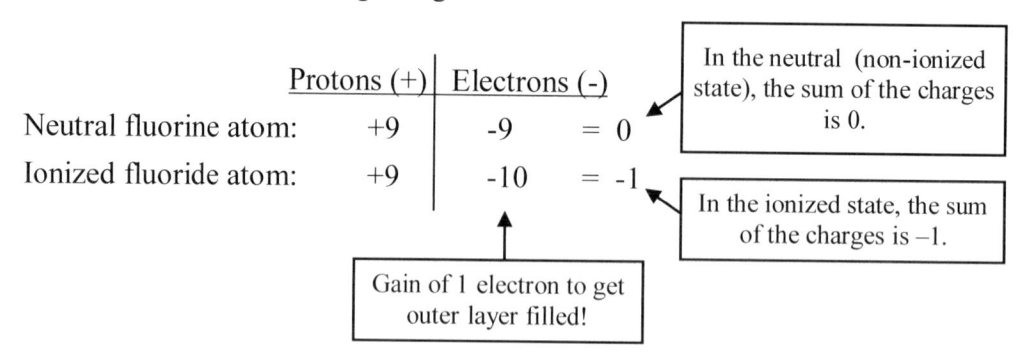

	Protons (+)	Electrons (-)	
Neutral fluorine atom:	+9	-9	= 0
Ionized fluoride atom:	+9	-10	= -1

In the neutral (non-ionized state), the sum of the charges is 0.

In the ionized state, the sum of the charges is –1.

Gain of 1 electron to get outer layer filled!

$$\text{Fluoride} = F^{-1}$$

Let's look at another example:

In the neutral, non-ionized state, the chlorine atom has 17 positively charged protons and 17 negatively charged electrons. Recall that chlorine is a halogen and desires to become more stable by gaining one electron to take on the appearance of its neighboring noble gas family member, argon. The addition of one electron results in an excess negative charge of one.

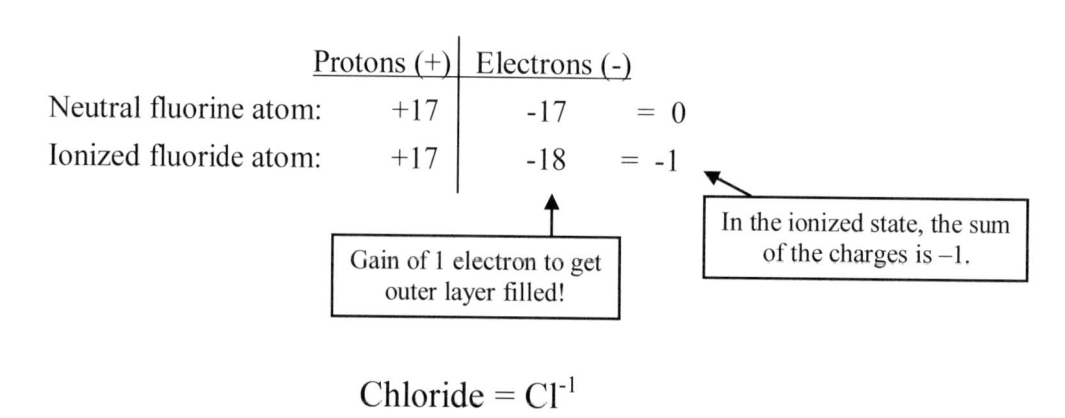

	Protons (+)	Electrons (-)	
Neutral fluorine atom:	+17	-17	= 0
Ionized fluoride atom:	+17	-18	= -1

Gain of 1 electron to get outer layer filled!

In the ionized state, the sum of the charges is –1.

$$\text{Chloride} = Cl^{-1}$$

Because each member of the halogen family desires to ionize by gaining one electron to take on the configuration of a neighboring noble gas family member, the resulting charge on all of the halogen family members after ionization is –1.

Fluorine ionizes to fluoride: F^{-1}
Chlorine ionizes to chloride: Cl^{-1}
Iodine ionizes to iodide: I^{-1}
Astatine ionizes to astatide: At^{-1}
Bromine ionizes to bromide: Br^{-1}

What about the oxygen family just to the left of the halogens? Can you guess the resulting charge when each oxygen family member ionizes? Let's take a look at oxygen..

In the non-ionized, neutral state, an atom of oxygen has 8 protons and 8 electrons (six valence electrons). To achieve the stability of its nearest noble gas family member (neon), an atom of oxygen must gain two electrons. This results in an excess of two negative charges. Oxygen, therefore, takes on a –2 charge as it ionizes to oxide!

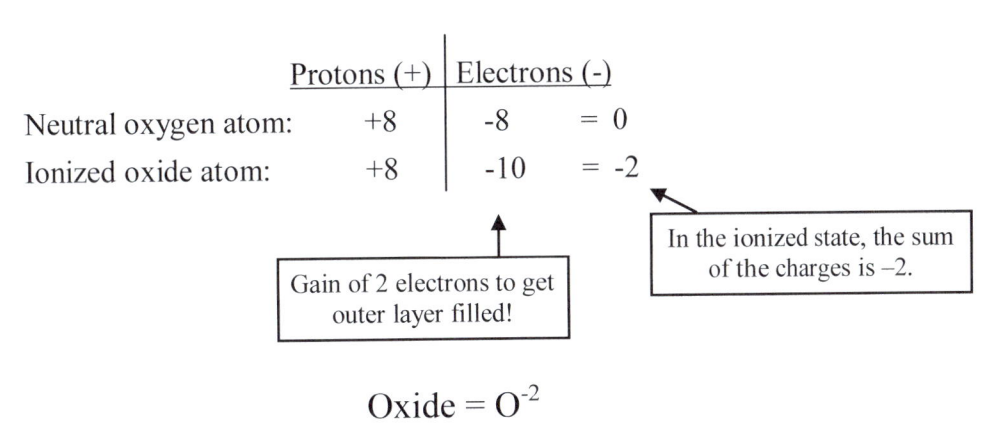

	Protons (+)	Electrons (-)	
Neutral oxygen atom:	+8	-8	= 0
Ionized oxide atom:	+8	-10	= -2

Gain of 2 electrons to get outer layer filled!

In the ionized state, the sum of the charges is –2.

$$Oxide = O^{-2}$$

As you might recall, the remaining members of the oxygen family also ionize by gaining two electrons to achieve the configuration of the nearest noble gas family member. Therefore, as each oxygen family member ionizes, the resulting charge for each ion will be –2.

Oxygen ionizes to oxide: O^{-2}
Sulfur ionizes to sulfide: S^{-2}
Selenium ionizes to selenide: S^{-2}
Tellurium ionizes to telluride: Te^{-2}
Polonium ionizes to polonide: Po^{-2}

At this point, let's turn our attention to the sodium family on the far left side of the periodic table of elements. Recall that each member of the highly reactive sodium family has one lone valence electron which each member would "love" to lose. By losing this single electron, each member of the sodium family would end up having an excess number of positive charges.

Take sodium for example. In the non-ionized or neutral state, the sodium ion has 11 positively charged protons and 11 negatively charged electrons (1 valence electron). To become more stable, the sodium ion loses that one valence electron to achieve the electron configuration of the nearest noble gas family member, neon. In doing so, the sodium atom ionizes to form the sodium ion which now has one more positive charge than negative. The result is a sodium ion with a +1 charge.

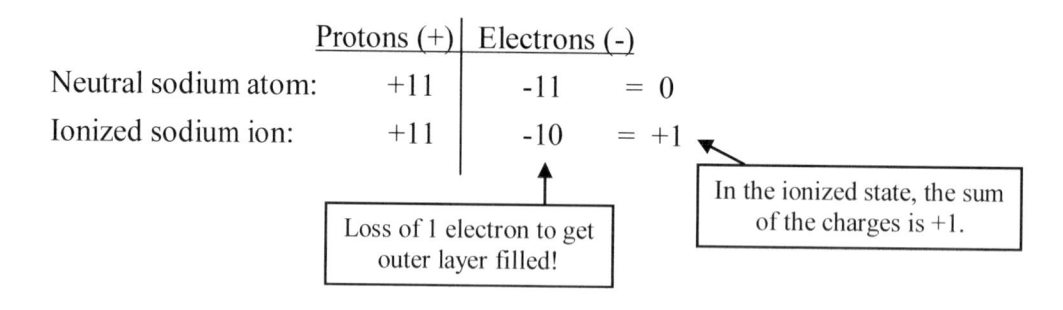

	Protons (+)	Electrons (-)	
Neutral sodium atom:	+11	-11	= 0
Ionized sodium ion:	+11	-10	= +1

Loss of 1 electron to get outer layer filled!

In the ionized state, the sum of the charges is +1.

Sodium ion: Na^{+1}

Each of the remaining members of the sodium family also ionizes by losing its one lone valence electron. As each family member ionizes, each atom acquires an excess of one positive charge which results in ions with a +1 charge.

Hydrogen ionizes to the hydrogen ion: H^{+1}
Lithium ionizes to the lithium ion: Li^{+1}
Sodium ionizes to the sodium ion: Na^{+1}
Potassium ionizes to the potassium ion: K^{+1}
Rubidium ionizes to the rubidium ion: Rb^{+1}
Cesium ionizes to the cesium ion: Cs^{+1}
Francium ionizes to the francium ion: Fr^{+1}

What do you suppose would be the resulting charge when members of the calcium family ionize? Recall that each member of the calcium family has two valence electrons. By losing those two valence electrons, members of the calcium family can ionize and achieve the stability found among the noble gas family members. Losing those two electrons, each having negative charges, will result in an excess of two positive charges. The resulting ions will have a +2 charge.

For example, let's look at the calcium family member magnesium. Magnesium, atomic number 12, has a total of 12 positively charged protons and 12 negatively charged electrons. In the process of ionizing, an atom of magnesium will lose its two outermost electrons (valence electrons) to take on the configuration of noble gas family member neon (atomic number 10). The atom of magnesium will now have 12 protons and 10 electrons—an excess of 2 protons!

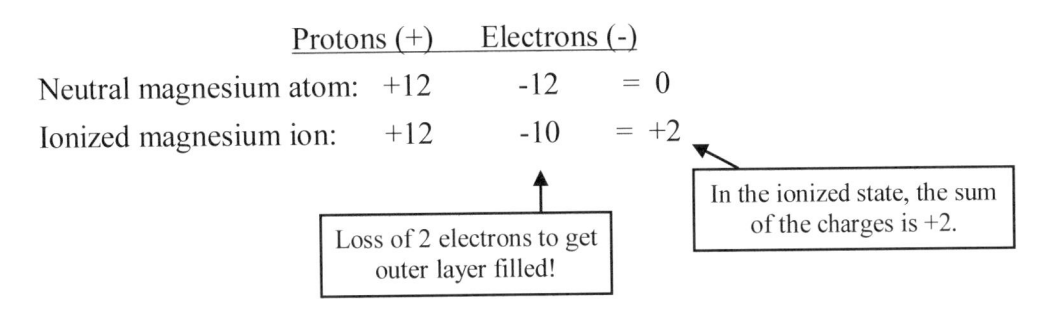

```
                        Protons (+)    Electrons (-)
Neutral magnesium atom:    +12            -12       = 0
Ionized magnesium ion:     +12            -10       = +2
```

Loss of 2 electrons to get outer layer filled!

In the ionized state, the sum of the charges is +2.

Magnesium ion: Mg^{+2}

The remaining members of the calcium family acquire a +2 charge as they ionize.

Beryllium ionizes to the beryllium ion: Be^{+2}
Magnesium ionizes to the magnesium ion: Mg^{+2}
Calcium ionizes to the calcium ion: Ca^{+2}
Strontium ionizes to the strontium ion: Sr^{+2}
Barium ionizes to the barium ion: Ba^{+2}
Radium ionizes to the radium ion: Ra^{+2}

Let's pause to review what we have discussed in this chapter. We explored the idea that elements, which are naturally unstable in their neutral form, gain or release electrons to acquire more stable electron configurations similar to noble gas members. The number of electrons gained or lost determines the resulting charge of the ion. We discussed that members of the halogen family acquire a –1 charge as they ionize and the endings of their names change from –ine to –ide (fluorine to fluoride, for example). Members of the oxygen family acquire a –2 charge as they ionize and they, too, change the ending of their names to –ide (oxygen to oxide). Members of the sodium family acquire a +1 charge while members of the calcium family acquire a +2 charge. The names of the positively charged ions do not change, but the word "ion" is added.

One last concept to consider is that ions which have a positive charge are referred to as **cations** (pronounced either CAT-ions or KAY-shuns as in nations). Ions which have a negative charge are referred to as anions (pronounced AN-i-uns). Ions formed by members of the halogen and oxygen families are anions. Ions formed by members of the sodium and calcium families are cations.

Practice learning the names, symbols and charges of these ions by playing Ion Bingo. Your teacher will have instructions for creating your bingo card and how to play. A list of ions to use to fill your card is found below.

Ion Name	Ion Symbol	Ion Name	Ion Symbol
Hydrogen Cation	H^{+1}	Oxide	O^{-2}
Lithium cation	Li^{+1}	Sulfide	S^{-2}
Sodium cation	Na^{+1}	Selenide	Se^{-2}
Potassium cation	K^{+1}	Telluride	Te^{-2}
Rubidium cation	Rb^{+1}	Polonide	Po^{-2}
Cesium cation	Cs^{+1}	Fluoride	F^{-1}
Francium cation	Fr^{+1}	Chloride	Cl^{-1}
Beryllium cation	Be^{+2}	Bromide	Br^{-1}
Magnesium cation	Mg^{+2}	Iodide	I^{-1}
Calcium cation	Ca^{+2}	Astatide	At^{-1}
Strontium cation	Sr^{+2}	Radium	Ra^{+2}
Barium cation	Ba^{+2}		

Use this list of monoatomic ions to create your Ion Bingo card.

Friendly Chemistry

Ion Bingo!

Write the symbol and charge of ions in the spaces below (one ion per space). Your teacher will call out ion names. Place a marker on the ions called. Get five in a row across, down or diagonally and you've got a BINGO!

Friendly Chemistry

Ion Bingo!

Write the symbol and charge of ions in the spaces below (one ion per space). Your teacher will call out ion names. Place a marker on the ions called. Get five in a row across, down or diagonally and you've got a BINGO!

Friendly Chemistry

Name_____ Date_____

Friendly Chemistry

Lesson 12
Ion Formation Practice

Below is a table which will help you practice the ionization process of the sodium, calcium, oxygen and halogen families of elements. The first problem has been completed for you.

The element	Of the ___ family	With ___ total e's	And ___ valence e's	Gains or loses	___ e's as it ionizes	To have the symbol and charge of ___	And the ion name of ___.
sodium	sodium	11	1	loses	1	Na^{+1}	Sodium cation
lithium							
potassium							
calcium							
barium							
hydrogen							
chlorine							
oxygen							
iodine							
cesium							
fluorine							
sulfur							
selenium							
polonium							
francium							
magnesium							
tellurium							
astatine							
strontium							

S115

Friendly Chemistry

Name_____ Date_____

Friendly Chemistry

Lesson 12 Ion Practice 2

Below is a list of ions. Tell first, the ion's symbol and charge. Then tell if the ion will be an anion or cation. The first problem is completed for you!

Ion Name	Ion Symbol and Charge	Is this an Anion or Cation?
1. Chloride	Cl^{-1}	Anion
2. Fluoride		
3. Sodium cation		
4. Iodide		
5. Magnesium cation		
6. Calcium cation		
7. Strontium cation		
8. Oxide		
9. Sulfide		
10. Hydrogen cation		
11. Beryllium cation		
12. Potassium cation		
13. Selenide		
14. Rubidium cation		
15. Lithium cation		

Lesson 13: Forming Compounds from Ions

In the last lesson, we discussed how members of certain families of the periodic table gain or lose electrons (ionize), to achieve more stable electron configurations. We discussed that these family members "strive" to acquire a configuration like those of the members of the noble gas family. We learned that the sodium and calcium family members desire to lose one or two electrons, respectively, and that the halogens desire to gain one electron and the oxygen family members want to gain two electrons.

With all of this in mind, it may seem logical that members of these different families might be able work out some sort of deal where family members exchange electrons for the benefit of both ions involved. In other words, electrons from atoms which desire to *give away* electrons can be received by atoms which desire to *gain* electrons. By doing so, chemical compounds can be formed.

Let's look at an example. Sodium (a member of the sodium family) has a strong desire to ionize by getting rid of its lone valence electron. On the other hand, chlorine, a halogen family member, has a desire to gain one electron to gain the stability of its neighboring noble gas family member, argon. If sodium cations are mixed with chlorine anions, the sodium cations readily "donate" their lone valence electrons to the chlorine anions and in the process create a "relationship" known as an ionic bond. A **bond** is like a linkage or attachment between atoms. As long as sodium ions have access to chlorine ions they will form **ionic bonds** and, together, they are quite stable.

In fact, the bond created between sodium and chlorine is so stable that the once extremely reactive (even explosive!) sodium atoms and highly poisonous chlorine atoms together create a **compound** ingested on a daily basis by persons all around the world: table salt. Yes, table salt, with the chemical name of sodium chloride, is a very stable compound very much *unlike* its two ingredient elements. The formation of the ionic bond between sodium and chloride results in two "happy" ions and a very stable and relatively safe "relationship."

Note that in the sodium chloride compound, it only took one atom of each ingredient to create the stable compound. We can write this newly formed compound using element symbols, superscripts and subscripts. We will utilize superscripts to indicate the charge on each ion and subscripts to tell how many of each ion were needed to create the stable compound. Here is how you would write the formula for sodium chloride:

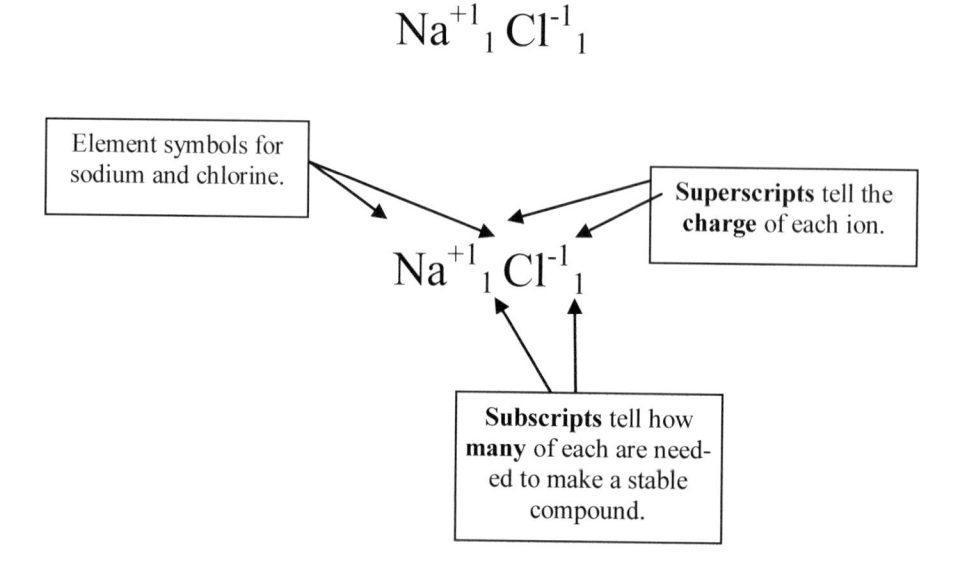

$$Na^{+1}_{1} Cl^{-1}_{1}$$

Element symbols for sodium and chlorine.

Superscripts tell the **charge** of each ion.

$$Na^{+1}_{1} Cl^{-1}_{1}$$

Subscripts tell how **many** of each are needed to make a stable compound.

Note, also, that if you were to add the charges (superscripts), the sum would be zero (no net charge). This tells us we have accurately written the formula for the chemical compound.

$$Na^{+1}_{1} Cl^{-1}_{1} = 0$$

$$+1 + (-1) = 0$$

The sum of the charges must equal zero. Always check your work by making sure that the sum of the charges always equals zero!

Let's look at another example. If we took some potassium and some fluorine atoms, could we create a stable compound? Recall that potassium is a member of the sodium family and each member of that family has one lone valence electron. Recall, also, that fluorine is a halogen and is in need of one electron to completely fill its outermost orbit (which would allow it to acquire the configuration of neon). Potassium could fulfill the needs of fluorine by supplying the needed electron and, *viola*, an ionic bond is formed between the potassium cation and the fluoride anion. Together, the two ions create a stable, ionic compound known as potassium fluoride. Note that it only took one ion of each atom to create this compound. In writing the compound formula, we will use 1's for subscripts to indicate how many of each ion we used.

$$K^{+1}_{1} F^{-1}_{1}$$

Note, once again, that the charges on each ion if added together equal zero indicating that we have accurately written the formula for the compound.

$$K^{+1}_{1} F^{-1}_{1} = 0$$

As you can guess, one atom of each member of the sodium family can join with one atom of each member of the halogen family. The sodium family members desire to lose one electron and the halogen family member desire to gain one electron. Therefore, relatively stable compounds can be created by combining members of the sodium family with members of the halogen family.

S119

Now, let's look at members of the oxygen family. Recall that each member of the oxygen family desires to *gain* **two** electrons to achieve the stability of members of the noble gas family (oxygen family members have 6 valence electrons). Can you find another family of elements on the periodic table which has a desire to *lose* two electrons? If you chose the calcium family, you are correct! The members of the calcium family, as you recall, have two lone electrons in their outermost orbits and desire to lose those two electrons to achieve the stability of members of the noble gas family. Together, oxygen family members (desiring to gain two electrons) and calcium family members (desiring to lose two electrons) can make stable ionic compounds.

Let's look at some examples. Magnesium is a member of the calcium family and desires to lose two electrons. Sulfur is a member of the oxygen family and desires to gain two electrons. Together they create the stable compound magnesium sulfide.

$$Mg^{+2}_{\ 1} S^{2-}_{\ 1}$$

Note that to create magnesium sulfide, one atom of magnesium (subscript of 1) and one atom of the sulfide ion (subscript of 1) are required. Note, also, that once again if you add the charges of each ion together, the sum is zero.

$$Mg^{+2}_{\ 1} S^{-2}_{\ 1} = 0$$

Let's look at another example: calcium and oxygen. Calcium has the desire to lose its two valence electrons and oxygen has the desire to gain two additional electrons to completely fill its outermost orbit (energy level). Together, they can create a stable compound known as calcium oxide.

$$Ca^{+2}_{\ 1} O^{-2}_{\ 1}$$

Note that it takes one calcium ion for every one oxide ion to create this compound (subscripts of 1). Again, the charges (superscripts), when added, equal zero.

$$Ca^{+2}_{\ 1} O^{-2}_{\ 1} = 0$$

The remaining members of both the calcium and oxygen families can form compounds in the same way (one calcium family member for each oxygen family member).

At this point you might be wondering, "What happens when a member of the sodium family forms a compound with a member of the oxygen family? The sodium family member wants to give away **one** electron, but the oxygen family member wants to gain **two** electrons."

Let's take a look at an example of this situation: sodium oxide. The sodium cation desires to lose its one valence electron. The oxide anion has the desire to gain two electrons in order to achieve the stability of neon. In order for the compound to form, it will take **two** sodium cations each contributing one valence electron, for every **one** oxide anion. Together, all three of these ions can form the stable compound sodium oxide.

$$Na^{+1}_2 \ O^{-2}_1$$

> Note that we wrote a subscript of 2 this time to indicate we needed **two** sodium cations to meet the needs of the oxide anion.

Recall that to correctly write a stable compound, the total of the charges must equal zero. In this case for the sodium cation, we have 2 times +1 = +2. For the oxide we have 1 times (-)2 = (-)2. If we add these together, +2 + (-)2, we get the sum of zero! Therefore, we can say in order to write the correct compound name for sodium oxide, it requires we use **two** sodium cations for every **one** oxide anion.

Let's look at one more example: calcium chloride. Recall that calcium has **two** valence electrons it would like to "donate" to some other anion. In this case, the chloride anions each only want to gain **one** electron. How can the calcium chloride compound form? Maybe you can see that **two** chloride anions can meet the needs of each calcium cation. The calcium cation "gives" one electron to one chloride anion and "gives" the second electron to the other chloride anion. Together, the three ions can form the stable compound of calcium chloride. The compound will look like this:

$$Ca^{+2}_1 \ Cl^{-1}_2$$

> Note that we wrote a subscript of 2 this time to indicate we needed **two** chloride anons to meet the needs of the one calcium cation.

Recall that to correctly write a stable compound, the total of the charges must equal zero. In this case for the calcium cation, we have 1 times +2 = +2. For the chloride we have 2 times (-)1 = (-)2. If we add these together, +2 + (-)2, we get the sum of zero! Therefore, we can say in order to write the correct compound name for calcium chloride, it requires we use **two** chloride anions for every **one** calcium cation.

Let's review what we have discussed in this lesson. We stated that atoms have a desire to become more stable by ionizing (gaining or losing electrons). Atoms that desire to **lose** electrons can donate those valence electrons to atoms which desire to **gain** electrons. In the process of transferring these valence electrons from one atom to another, an ionic bond is formed and the two ions are linked to form a compound. We explored how members of the sodium and halogen families form compounds and how members of the calcium and oxygen families form compounds (all on a one-to-one ratio of cation to anion). We also looked at how compounds can form between ions where it may require more of one ion than the other to form the compound (not a one-to-one ratio). Finally, we checked our work by making sure the sum of the charges always added to zero.

Compounds formed by ionic bonding that you may be familiar with are sodium chloride ($Na^{+1+}{}_1Cl^{-1}{}_1$; table salt) and sodium fluoride ($Na^{+1+}{}_1F^{-1}{}_1$; the fluoride compound found in toothpaste). Hydrogen chloride ($H^{+1+}{}_1Cl^{-1}{}_1$; or, as it is known to chemists, hydrochloric acid), is the "acid" found in your stomach which breaks down food.

One last thing to realize when writing compounds is the fact that the **cation** portion of the compound is always written first followed by the **anion** portion.

Practice all of these new ideas by completing the practice pages which follow!

Name _____ Date _____

Friendly Chemistry

Lesson 13
Writing Simple Chemical Formulas –1

In the first column of the table below you will see the name of a simple chemical compound. Write its chemical formula in the second column. Make certain that the charges add to zero!

Chemical Name	Chemical Formula
1. Sodium fluoride	
2. Potassium chloride	
3. Lithium bromide	
4. Hydrogen chloride	
5. Magnesium oxide	
6. Calcium sulfide	
7. Beryllium oxide	
8. Potassium iodide	
9. Cesium chloride	
10. Sodium bromide	

Name _____ Date _____

Friendly Chemistry

Lesson 13
Writing Simple Chemical Formulas—2

In the first column of the table below you will see the name of a simple chemical compound. Write its chemical formula in the second column. Make certain that the charges add to zero!

Chemical Name	Chemical Formula
1. Sodium iodide	
2. Potassium chloride	
3. Lithium oxide	
4. Hydrogen sulfide	
5. Magnesium bromide	
6. Calcium sulfide	
7. Beryllium polonide	
8. Strontium iodide	
9. Cesium chloride	
10. Sodium selenide	

Lesson 14: Learning (some slightly) More Complex Ions

In the last lesson, we discussed how atoms form compounds by transferring or sharing valence electrons. We discussed the idea that atoms form compounds in order to acquire a more stable electron configuration. In this lesson, we will explore ions which are somewhat more complex than those we examined previously which also join to form compounds. Let's get started!

The ions we examined in the last lesson (both positively charged cations and negatively charged anions) consisted of single atoms. Each ion only had one type of element present. These ions are known as **monatomic ions** (mono- meaning one). The ions we will explore in this lesson are made up of two or more types of elements. Ions made up of more than one type of element are known as **polyatomic ions** (poly- meaning many). We will also examine ions which have more than one charge. In some chemistry texts you may find that instead of the word "charge", the word "oxidation state" is used. Both terms are correct. We will continue to use the word "charge" in our discussions.

Let's begin by discussing some features of polyatomic ions. Like the monatomic ions, polyatomic ions can be cations (positively charged) or anions (negatively charged). It is important to realize that even though the polyatomic ions are made up of two or more elements, the ion acts as one unit and the charge for the polyatomic ion is for the whole unit. When writing the formula for a polyatomic ion, it is customary to enclose the element symbols within parentheses and place the charge outside the parentheses to indicate the charge is for the ion as a whole (and not for any individual component within the polyatomic ion). Look at these examples.

Nitrate is a polyatomic anion made up of **one nitrogen** atom and **three oxygen** atoms. Its formula is $(N_1O_3)^{-1}$.

> Note that the –1 charge is for the entire nitrate ion and *not* for any individual portion of the ion.

Hydroxide is a polyatomic anion made up of **one hydrogen** atom and **one oxygen** atom. Its formula is $(O_1H_1)^{-1}$.

> Note that the –1 charge is for the entire hydroxide ion and *not* for any individual portion of the ion.

Ammonium is a polyatomic cation made up of **one nitrogen** atom and **four hydrogen** atoms. Its formula is $(N_1H_4)^{+1}$.

> Note that the +1 charge is for the entire ammonium ion and *not* for any individual portion of the ion.

Hydronium is a polyatomic cation composed of **three hydrogen** atoms and **one oxygen** atom. Its formula is $(H_3O)^{+1}$.

> Note that the +1 charge is for the entire hydronium ion and *not* for any individual portion of the ion.

The names of the polyatomic ions may seem confusing at first, but there are some keys which can give you clues to what ions are present in the polyatomic ion.

1. Polyatomic ion names which have the ending **"-ite" or "-ate"** always have **oxygen** present. For example, nitr<u>ite</u> is composed of nitrogen **and oxygen**. Sulf<u>ate</u> is composed of sulfur **and oxygen**.

2. In addition, the ions which end in **"-ite"** always have **one less oxygen** than those ions ending in "-ate". For example, nitr<u>ite</u> has two atoms of oxygen where nitr<u>ate</u> has three atoms of oxygen. Sulf<u>ite</u> has three atoms of oxygen where sulf<u>ate</u> has four atoms of oxygen.

3. Ions with the –ide, -ite or –ate endings are always **negatively** charged.

4. The prefix **hypo–** means **low**. Hypochlorite as a very low amount of oxygen atoms present.

5. The prefix **per-** (which can be considered a shortened form of hyper-) means **high** or excess. Perchlorate or permanganate each carry a "heavier" load of oxygen atoms (both having four oxygen atoms each).

6. For certain ions having more than one charge, it is customary to used Roman Numerals to designate the charge intended for the ion in the name of compound. For example, the element mercury can have two possible charges: +1 and +2. When writing the name for mercury with the +1 charge, we write mercury (I) and when writing the name for mercury with the +2 charge, we write mercury (II). See your Master Ion list for additional ions that utilize Roman Numerals to indicate their charge (lower right column). Note that these ions are always positively charged ions (cations)!

Plus, here are some memory "helpers" to assist you in learning the symbols and charges for these ions.

1. For the ions sulfate and sulfite: "We get out of school between 3 and 4, but sometimes at 2 on Fridays." Sulfite has 3 oxygens, sulfate has 4 while the charge for both ions is a –2. The letter "s" of school links this mnemonic rule to the sulfates.

Friendly Chemistry

2. For the ions nitrite and nitrate: It's night between 2 and 3, but darker at 1. The term night links this mnemonic to the nitrite/ate set of ions. Nitrite has two oxygens while nitrate has 3 oxygen atoms. Both of them have a −1 charge.

3. To help remember the charges for the cations chromium and aluminum: Chromium and Aluminum are 3rd cousins. Both have a +3 charge.

4. For all chlorine containing anions: It's hot around the pool at 1. All chlorine containing anions have a negative one charge. Chlorine is used in swimming pool water.

5. For the anion, phosphate: It takes four matches to make a fire, but three will work. Phosphate is used in making matches and has four oxygen atoms and a −3 charge.

6. For the hydronium cation: Hydronium is glorified water. Water has two hydrogen atoms, but hydronium is glorified by having three hydrogen atoms!

7. For the nickel cation: The "N" symbol looks like the number two (only on its side). This is also useful for the zinc cation: The "Z" looks like a two. Both cations have +2 charges.

8. For the chromate and dichromate anions: The chromates are relatives of the sulfates—both having −2 charges.

9. For the carbonate ion: If you drink a Coke at 3, you'll be up until 2. Carbonate is an ingredient used to make carbonated water found in pop. Carbonate has three oxygen atoms and a −2 charge.

10. For the hydroxide ion: "Oh, it's a hydrox cookie." The "oh" is the symbol for the hydroxide ion. (Hydrox cookies are a form of Oreos available in the Midwest in case this memory aid makes no sense to you!)

Friendly Chemistry

Name_____ Date_____

Friendly Chemistry

Master List of Ions

Anions (−) Cations (+)

Anion	Formula	Cation	Formula
Acetate	$(C_2H_3O_2)^{-1}$	Aluminum	Al^{+3}
Bromide	Br^{-1}	Chromium	Cr^{+3}
Carbonate	$(CO_3)^{-2}$	Hydrogen	H^{+1}
Chloride	Cl^{-1}	Sodium	Na^{+1}
Hypochlorite	$(ClO_1)^{-1}$	Lithium	Li^{+1}
Chlorite	$(ClO_2)^{-1}$	Potassium	K^{+1}
Chlorate	$(ClO_3)^{-1}$	Calcium	Ca^{+2}
Perchlorate	$(ClO_4)^{-1}$	Magnesium	Mg^{+2}
Fluoride	F^{-1}	Strontium	Sr^{+2}
Iodide	I^{-1}	Nickel	Ni^{+2}
Oxide	O^{-2}	Zinc	Zn^{+2}
Sulfide	S^{-2}	Silver	Ag^{+1}
Sulfite	$(SO_3)^{-2}$	Hydronium	$(H_3O)^{+1}$
Sulfate	$(SO_4)^{-2}$	Cesium	Cs^{+1}
Nitrite	$(NO_2)^{-1}$	Ammonium	$(NH_4)^{+1}$
Nitrate	$(NO_3)^{-1}$	Mercury (I)	Hg^{+1}
Chromate	$(CrO_4)^{-2}$	Mercury (II)	Hg^{+2}
Dichromate	$(Cr_2O_7)^{-2}$	Copper (I)	Cu^{+1}
Cyanide	$(CN)^{-1}$	Copper (II)	Cu^{+2}
Permanganate	$(MnO_4)^{-1}$	Tin (II)	Sn^{+2}
Peroxide	$(O_2)^{-2}$	Tin (IV)	Sn^{+4}
Phosphate	$(PO_4)^{-3}$	Lead (II), (IV)	$Pb^{+2}\ Pb^{+4}$
Hydroxide	$(OH)^{-1}$	Iron (II), (III)	$Fe^{+2}\ Fe^{+3}$

Friendly Chemistry

Polyatomic Ion Bingo!

Write the ion symbols and charges in the spaces below. Include one "free space!" As the ion names are called place a marker on the appropriate ion symbol. Call "bingo" when you get six across, six down or six diagonally and then tell the ions you used to make the bingo to win your prize!!!

NH_4^{+1}					

Name_____ Date_____

Friendly Chemistry

Lesson 14: Ion Practice - 1

Below is a chart with the names of several of the ions we have been studying. Write the correct symbol and charge or name for these ions. Use your master sheet to make sure you are writing the correct symbols, charges or names!

Ion name	Ion symbol and charge	Ion symbol and charge	Ion name
1. Calcium cation		19. $(C_2H_3O_2)^{-1}$	
2. Sodium cation		20. I^{-1}	
3. Fluoride		21. $(NO_2)^{-1}$	
4. Chloride		22. $(CN)^{-1}$	
5. Hypochlorite		23. Ni^{+2}	
6. Chlorite		24. Sn^{+2}	
7. Perchlorate		25. Ag^{+1}	
8. Sulfite		26. S^{-2}	
9. Sulfate		27. Sn^{+4}	
10. Nitrate		28. $(ClO_1)^{-1}$	
11. Nitrite		29. $(SO_4)^{-2}$	
12. Hydroxide		30. $(ClO_3)^{-1}$	
13. Zinc		31. $(ClO_4)^{-1}$	
14. Hydrogen cation		32. $(ClO_2)^{-1}$	
15. Chromate		33. $(ClO_1)^{-1}$	
16. Dichromate		34. $(OH)^{-1}$	
17. Silver cation		35. Li^{+1}	
18. Nickel cation		36. Ba^{+2}	

Friendly Chemistry

Name_____ Date_____

Friendly Chemistry

Lesson 14: Ion Practice - 2

Complete the chart below by writing the appropriate ion symbol and charge or ion name.

Ion name	Ion symbol and charge	Ion symbol and charge	Ion name
1. sulfate		17. $(ClO_3)^{-1}$	
2. sulfite		18. Br^{-1}	
3. oxide		19. Hg^{+2}	
4. perchlorate		20. $(NO_3)^{-1}$	
5. iodide		21. $(PO_4)^{-3}$	
6. Lithium cation		22. $(OH)^{-1}$	
7. Sodium cation		23. $(ClO_4)^{-1}$	
8. Lead (II)		24. $(MnO_4)^{-1}$	
9. phosphate		25. H^{+1}	
10. sulfide		26. $(ClO_2)^{-1}$	
11. hypochlorite		27. Cl^{-1}	
12. ammonium		28. $(SO_3)^{-2}$	
13. Calcium cation		29. $(CrO_4)^{-2}$	
14. Potassium cation		30. $(NH_4)^{+1}$	
15. Barium cation		31. $(O_2)^{-2}$	
16. hydroxide		32. Na^{+1}	

S132

Friendly Chemistry

Name_____ Date_____

Friendly Chemistry

Lesson 14: Ion Practice –3

Complete the chart below by writing the appropriate ion symbol and charge or ion name.

Ion name	Ion symbol and charge	Ion symbol and charge	Ion name
1. Chlorite		19. Sn^{+4}	
2. Sodium cation		20. I^{-1}	
3. Sulfite		21. $(ClO_1)^{-1}$	
4. Chloride		22. $(CN)^{-1}$	
5. Nitrate		23. $(SO_4)^{-2}$	
6. Calcium cation		24. Sn^{+2}	
7. Perchlorate		25. Ag^{+1}	
8. Zinc		26. S^{-2}	
9. Sulfate		27. $(C_2H_3O_2)^{-1}$	
10. Hypochlorite		28. $(ClO_1)^{-1}$	
11. Nitrite		29. Ni^{+2}	
12. Hydroxide		30. $(ClO_3)^{-1}$	
13. Fluoride		31. $(ClO_4)^{-1}$	
14. Hydrogen cation		32. Na^{+1}	
15. Nickel cation		33. $(NO_2)^{-1}$	
16. Dichromate		34. $(OH)^{-1}$	
17. Silver cation		35. $(ClO_2)^{-1}$	
18. Chromate		36. Sr^{+2}	

S133

Lesson 15: Making Compounds with Polyatomic Ions

Forming compounds with polyatomic ions is very similar to the way monoatomic ions form compounds. You must remember that the polyatomic ions (although made of more than one individual element) function as one unit. The charge given to the polyatomic ion is for the whole ion and not any individual portion. Let's look at some examples of how compounds made with polyatomic ions form.

Sodium hydroxide is a compound commonly known as lye which is used to unclog drains and to make lye soap. The ions which make sodium hydroxide are the cation sodium and the anion hydroxide.

Sodium   **Hydroxide**
Na^{+1} $(OH)^{-1}$

To make the compound sodium hydroxide, we would need to use one sodium ion and one hydroxide ion. In doing so, the sum of the charges equals zero. Recall, also, that it is customary to write the cation first, followed by the anion.

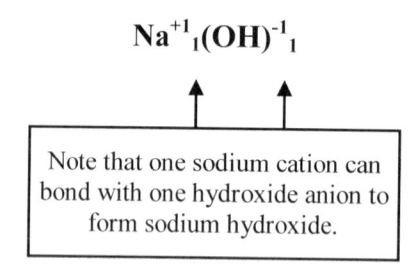

$$Na^{+1}_{1}(OH)^{-1}_{1}$$

> Note that one sodium cation can bond with one hydroxide anion to form sodium hydroxide.

Let's look at some more examples. **Calcium carbonate** is the chemical which makes up the shells of many ocean creatures. Another name for fossilized calcium carbonate is limestone. The constituents of calcium carbonate are the calcium cation and the carbonate anion.

$$\textbf{Calcium} \qquad \textbf{Carbonate}$$
$$Ca^{+2} \qquad (CO_3)^{-2}$$

In order to make the compound calcium carbonate, the charges of the ions must add up to zero. In this case, using one atom of each ion will result in a total charge of zero.

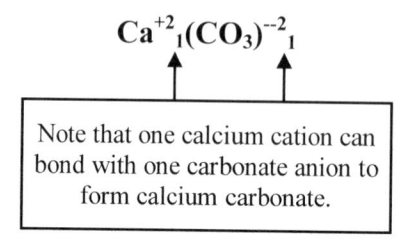

$$Ca^{+2}_{1}(CO_3)^{-2}_{1}$$

> Note that one calcium cation can bond with one carbonate anion to form calcium carbonate.

Here is another example. **Hydrogen peroxide** is a well-known antiseptic solution used to "bubble-out" impurities in shallow wounds. The hydrogen peroxide compound is composed of the hydrogen cation and the peroxide anion.

$$\textbf{Hydrogen} \qquad \textbf{Peroxide}$$
$$H^{+1} \qquad (O_2)^{-2}$$

If we take one hydrogen cation and one peroxide ion, the sum of the charges does *not* add up to zero. We must take a multiple of one or both of the ions in order for the sum of the charges to equal zero. In this case, taking two hydrogen cations (each with a +1 charge) and one peroxide anion (with its -2 charge) will result in a total charge of zero.

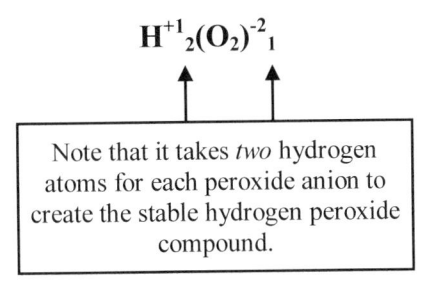

$$H^{+1}_2(O_2)^{-2}_1$$

> Note that it takes *two* hydrogen atoms for each peroxide anion to create the stable hydrogen peroxide compound.

And here is a final example. The chemical formula for rust is **iron (III) oxide**. The iron (III) oxide compound is composed of the iron (III) cation and the oxide anion. Recall that ions with a Roman Numeral as part of their name, tell you the charge of the cation. In this case iron (III) is Fe^{+3}.

$$\text{Iron (III)} \quad \text{Oxide}$$
$$Fe^{+3} \quad O^{-2}$$

Forming this compound is a little more challenging. In order to get our charges to add to zero, we must take **two** iron (III) cations and **three** oxide cations.

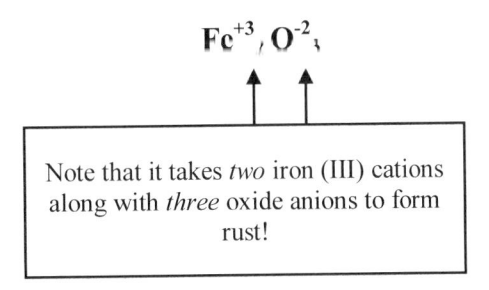

$$Fe^{+3}_2 O^{-2}_3$$

> Note that it takes *two* iron (III) cations along with *three* oxide anions to form rust!

Practice making compounds by completing the practice page which follows.

Name_____   Date_____

Friendly Chemistry

Lesson 15: Compound Writing Practice

Below you will see a list of compounds. Write the chemical formula for each compound. Refer to your Master Ion List or flashcards if necessary.

Compound Name	Chemical Formula	Compound Name	Chemical Formula
1. Sodium chloride		21. Barium hydroxide	
2. Potassium nitrate		22. Calcium nitrite	
3. Lithium chloride		23. Cesium sulfate	
4. Barium sulfite		24. Nickel nitrate	
5. Calcium carbonate		25. Calcium oxide	
6. Strontium oxide		26. Sodium bromide	
7. Tin (IV) sulfate		27. Silver chromate	
8. Hydrogen chlorate		28. Copper (II) fluoride	
9. Potassium perchlorate		29. Potassium permanganate	
10. Aluminum acetate		30. Lithium carbonate	
11. Magnesium sulfite		31. Nickel cyanide	
12. Mercury (II) bromide		32. Lead (II) hypochlorite	
13. Sodium dichromate		33. Zinc chlorite	
14. Strontium iodide		34. Hydrogen peroxide	
15. Ammonium dichromate		35. Copper (I) sulfite	
16. Potassium chloride		36. Hydronium sulfide	
17. Chromium perchlorate		37. Nickel phosphate	
18. Calcium cyanide		38. Mercury (II) carbonate	
19. Barium chlorite		39. Tin (IV) sulfate	
20. Ammonium chloride		40. Potassium hydroxide	

Lesson 16: Introducing the Mole

Now that you have an idea of how polyatomic ions form compounds, let's move on to explore how you might actually go about preparing some of these compounds. Before we begin, please realize that we will be discussing the preparation of these compounds in very simplistic ways and the actual preparation of these compounds in a laboratory might be quite complex. We will be learning the basic theory behind preparing compounds and *not* the actual laboratory processes for making these compounds.

Before you could prepare something to eat in your kitchen (biscuits for example), you would have to know *what* ingredients are necessary and *how much* of each ingredient you might need. We have already discussed how you determine the ingredients you will need (ie. writing the chemical formula from the chemical name of the compound; sodium chloride has the formula $Na^{+1}_{\ 1}Cl^{-1}_{\ 1}$). However, we have yet to discuss how you know how *much* of each ingredient you will need to make a particular compound. In our example of preparing biscuits, you would need flour, salt, baking powder, butter or shortening and milk. However, altering the *amounts* of these ingredients could cause you to end up with pancakes and not biscuits!

You will recall from this and the previous lesson that as we made compounds from both monoatomic and polyatomic ions, we worked hard to make sure that each ion acquired a stable configuration (having its outermost orbit completely filled with 8 electrons). We could check to be sure that we had indeed written a correct compound when the charges of the ions added to zero. In some cases we had to take a multiple of one or both ions to create a stable compound. Look at this example to refresh your memory.

$$\text{Calcium chloride}$$
$$Ca^{+2}{}_1 Cl^{-1}{}_2$$

In this example of calcium chloride, two chloride ions are required to accept the two valence electrons that each calcium ion would like to donate. To make the stable compound, two chloride anions (subscript of two) are required for every one (subscript of one) calcium cation.

At this point, we could say that we need two chloride ions for every one calcium ion. However, if you were asked to prepare a container of calcium chloride it would be extremely difficult, if not impossible, to isolate one atom of calcium and two chloride atoms! Obviously to make it easier, we need to work with larger amounts of each ingredient, keeping in mind the correct ratio of each ingredient required to make our desired final product. In other words, we are multiplying our basic recipe many, many times in order to make a large enough batch with which to work!

If we just multiply our basic recipe for calcium chloride again and again, we will still be working with a ratio of one atom to two atoms. We might decide that $Ca^{+2}{}_{100,000} Cl^{-1}{}_{200,000}$ would be an adequate ratio to give the desired amount of product needed. However, we still have the problem of counting out individual atoms of each ingredient - which would still be considered impossible!

In most chemistry laboratories, quantities of chemicals to be used in the preparation of a compound are usually "weighed-out" on a balance or scale. The units used to measure these amounts are grams. In order to know how many grams of a particular "ingredient" you need, you must know some sort of relationship that must exist between **grams** of an ingredient and the number of **atoms** present in that amount. Don't despair, because that relationship has already been worked out by a chemist by the name of Amadeo Avogadro (1766-1856).

Avogadro worked out the relationship between the number of grams of an element necessary to represent a certain number of atoms. The number he calculated, known as **Avogadro's number**, tells us the number of atoms of any element required to make one **mole** of that element. (Note that in this case we are using Avogadro's number to indicate the number of atoms in a mole of a *single particular element*. Avogadro's number can also be used to tell the number of atoms present in a mole of a *compound* as well. This will be discussed in a later lesson.)

Now, you might be saying, "Wait a minute here, you said that we were finding a relationship between atoms and *grams*! Where did this unit 'mole' come into play?"

Well, it turns out that to make **one mole of any element it takes 6.02×10^{23} atoms (Avogadro's number).** That is, it takes 602,000,000,000,000,000,000,000 atoms to make one mole of any element. You might think of a mole as being an arbitrary amount - kind of like a cup or teaspoon, realizing that **the number of atoms required to make a mole of any element is the same, but the weight or mass of each mole is quite variable based upon properties of each element.** A cup of lead BB's would obviously have a different mass than a cup of plastic packing peanuts. Likewise a mole of one element will very likely have a different mass than a mole of another element, although the same number of atoms of each element will be present in each mole.

It is very important to remember that in a mole of *any* element, there are the same number of atoms (6.02×10^{23} atoms), but the mass amounts of one mole of any element varies greatly!

We said that one mole of any element is equal to 6.02×10^{23} atoms of that element (Avogadro's number). If we look at our earlier example of calcium chloride, we can see that to make calcium chloride required one part of calcium ion for every two parts of chloride ion: $Ca^{+2}_1Cl^{-1}_2$. Let's substitute Avogadro's number of each ingredient in place of the 1 and 2. In other words we would have:

$$Ca^{+2}_{6.02 \times 10^{23}} Cl^{-1}_{2(6 \times 10^{23})} \quad \text{or} \quad Ca^{+2}_{1\ mole} Cl^{-1}_{2\ moles}$$

In other words, we are saying that we have **one mole** of calcium ion for every **two moles** of chloride ion. However, you might say, "That still does not tell how many *grams* of each ingredient I need to prepare this compound!" You are correct! Chemists have determined how many grams of each element are required to equal one mole of that element. This value is known as the **atomic mass,** and atomic mass values can be found on most standard periodic tables of elements.

Examine your periodic table of elements. Note that the atomic mass value is the number that is *not* a whole number (remember, the whole number is the **atomic number** which tells us the number of protons, neutrons or electrons in an atom of that element). The atomic mass number is generally a decimal number and usually written directly above or below the element symbol. In the table shown here, the atomic mass value is written directly beneath the element symbol.

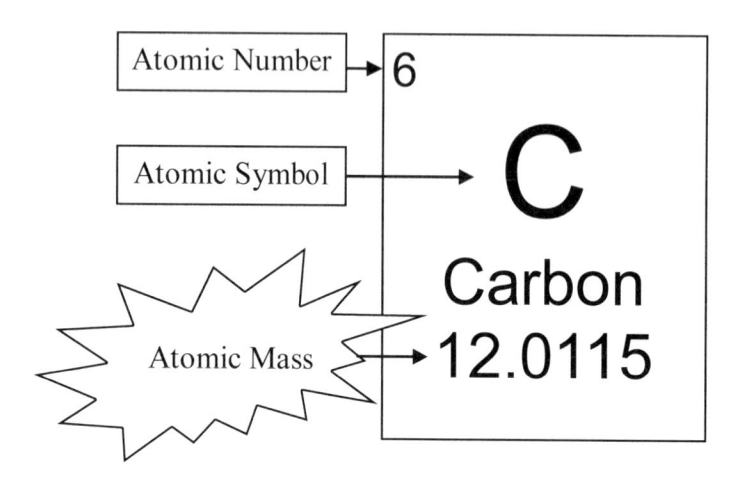

```
Atomic Number → 6
Atomic Symbol → C
                Carbon
Atomic Mass  → 12.0115
```

By using the atomic mass values from the periodic table, you can now "weigh-out" moles of chemicals. For example, if you needed one mole of carbon, by looking at the periodic table, you would know that you would need to weigh-out approximately 12 grams of carbon. If you needed two moles of carbon, you would just double this amount or approximately 24 grams of carbon.

Look at another example. Suppose you needed five moles of magnesium to do a lab activity. How many grams would you get from the storage container? To solve this problem, first find magnesium on the periodic table of elements. You will note that the atomic mass for magnesium is approximately 24 grams which means each mole of magnesium weighs about 24 grams. In order to get five moles of magnesium, you will need five times this amount or 5 X 24 grams/mole = 120 grams magnesium.

Let's review what you've learned in this lesson:

- A mole of any element has the same number of atoms which is 6.02×10^{23} atoms.
- Instead of saying the number of atoms or ions necessary to form a compound, we can now use moles of each compound constituent.
- Knowing about moles is useful because chemists know the mass of a mole of each element. This allows us to weigh-out moles of elements in the lab.

Practice making these mole to gram conversions on the next pages!

Friendly Chemistry

Name_____ Date_____

Friendly Chemistry

Lesson 16: Making Mole to Gram Conversions

Below are mole amounts of elements you need to get for your teacher. How many grams of each element do you need to weigh-out? Use your periodic table and calculator!

Moles Needed	Grams I need to weigh-out	Moles Needed	Grams I need to weigh-out
1. 2 moles of sodium	2 x 23 g/mole = 46 g	11. 5 moles of carbon	5 x 6 = 30
2. 10 moles of beryllium	10 x 4 = 40 g	12. 4.5 moles of lithium	4.5 x 3 = 13.5
3. 13 moles of boron	13 x 5 = 65 g	13. 7.5 moles of lead	7.5 x 82 = 615 g
4. 0.5 moles of cesium	0.5 x 55 = 27.5	14. 45 moles of barium	45 x 56 = 2520
5. 0.25 moles of helium	0.25 x 2 = 0.5	15. 14.75 moles of oxygen	14.75 x 8 = 118
6. 4.56 moles of calcium	4.56 x 20 = 41.2	16. 34 moles of silicon	34 x 14 = 476
7. 24.75 moles of aluminum	24.75 x 13 = 321.75 g	17. 23 moles of vanadium	23 x 23 = 529 g
8. 72 moles magnesium	72 x 12 = 864	18. 0.05 moles of gold	0.05 x 19 = 395
9. 34.7 moles of manganese	34.7 x 25 = 767.5	19. 4343.4 moles of sulfur	4343.4 x 16 = 69494.4
10. 0.15 moles of fluorine	0.15 x 9 = 1.35	20. 1 mole of nitrogen (a little easier being the last one!)	1 x 7 = 7 g

S143

Name_____ Date_____

Friendly Chemistry

Lesson 16: More Gram to Mole Conversion Practice

Below are problems that ask you to make gram to mole and mole to gram conversions. Take your time and work carefully. Ask questions if you're not sure what to do!

1. Mickey needed to get 5.5 moles of calcium from the store room. How many grams should Mickey get?

   220 g

2. Minnie was doing a chemical reaction which called for 4 moles of sodium metal. If she had a container which contained 100 grams of sodium metal and it was full, would she have enough? Support your answer with your calculations.

   92   no

3. Pluto had 5 moles of lithium. Yosemite Sam had 5 moles of carbon. Who had more atoms of their elements? Support your answer with calculations.

   5 x 7 = 35
   5 x 12 = 60   Sam

4. Fred had a recipe for making calcium carbonate. It called for 5 moles of carbon. If Barney had 55 grams of carbon, would he have enough to give Fred for the recipe? Support your answer with calculations.

   5 x 12 = 60   yes

S144

5. Daffy and Donald were looking at some labels on some chemicals. On one container it stated that 50 moles of magnesium were present in the container when new, however, the container had been partially used. If Daffy weighed the magnesium that remained the container and found it was 350 grams, how many moles were used?

1200
- 350
850

6. Sleepy told Dopey to get 56 moles of lead to make some fishing weights. Dopey returned with 11.6 kg of lead. Did he get enough to make the fishing weights? Support your answer with calculations.

207   11,592   yes

7. Hank and Drover were out chasing cows one day and the pickup ran out of gas. Drover had learned in his latest edition of Popular Science that corn could be make into ethanol which could be used like gas. In order for the recipe to work correctly, Drover needed 90 moles of oxygen gas. How many grams of oxygen should he get?

90 × 16 = 1440g

Lesson 17: Finding Formula Weights

In the last lesson, you learned that chemicals are measured in amounts called moles. You learned that a mole of any chemical contains 6.02×10^{23} atoms but the weight of a mole varies from element to element. For example, you learned that a mole of oxygen weighs about 16 grams while a mole of lead weighs about 208 grams. On your practice problems you were "asked" to weigh-out a specified number of moles of different chemicals. In order to do those problems, you referred to your periodic table of elements.

In this lesson, we will move one step further and apply the concepts we learned in the last lesson to finding amounts of *compounds*. This process is called finding a formula weight.

Let's begin by looking at the compound calcium chloride. Recall that in the compound calcium chloride, we had to take *two* chloride ions for every *one* calcium ion to achieve a stable compound.

$$Ca^{+2}_{1 \text{ mole}} Cl^{-1}_{2 \text{ moles}}$$

Or

$$Ca^{+2}_{1} Cl^{-1}_{2}$$

To find the formula weight, we examine all components of the compound. In this case, we have moles of calcium and chlorine. Begin by listing those components:

Ca:

Cl:

Next, write the number of moles of each component.

Ca: 1 mole

Cl: 2 moles

Next, multiply the number of moles present by the atomic mass of each element present.

Ca: 1 mole x 40 grams per mole =

Cl: 2 moles x 35 grams per mole =

Complete the arithmetic and then add the results together to find the formula weight.

Ca: 1 mole x 40.08 grams per mole = 40 grams
Cl: 2 moles x 35.453 grams per mole = 70 grams
Formula weight = 110 grams

We can then say that one mole of calcium chloride has the mass of 110 grams.

From the example above, you can see that the formula weight for a compound is the total number of grams found when the atomic masses of each ingredient are added. We can also say that the formula weight of a compound is equal to the mass of one mole of that compound. Knowing the formula weight of a compound will be quite useful as we continue to explore how to go about preparing compounds in the laboratory.

Before we continue, take some time to practice finding formula weights by working some practice problems. Below are three more examples.

Friendly Chemistry

Example 1. Find the formula weight of sodium hydroxide.

Step 1. Write the formula for the stable compound.

      Sodium: Na^{+1} **Hydroxide:** $(OH)^{-1}$

      Sodium Hydroxide: $Na^{+1}{}_1(OH)^{-1}{}_1$

We can say that to make one mole of sodium hydroxide, one mole of sodium ions and one mole of hydroxide ions are required.

Step 2. List the elements present in the compound.

Na:

O:

H:

Step 3. Find the number of moles present of each element and multiply by the atomic mass values for each element present.

Na: 1 x 23 grams = 23 grams

O: 1 x 16 grams = 16 grams

H: 1 x 1 gram = 1 gram

Step 4. Add up the grams.

Na: 1 x 23 grams = 23 grams
O: 1 x 16 grams = 16 grams
H: 1 x 1 gram = 1 gram
Formula weight = 40 grams

From this calculation we can say that one mole of sodium hydroxide would have the mass of 40 grams. If you needed one mole of sodium hydroxide, you would weigh-out 40 grams.

Friendly Chemistry

Example 2. Find the formula weight of potassium sulfate.

Step 1. Write the formula for the stable compound.

           Potassium: K^{+1}   **Sulfate: $(SO_4)^{-2}$**

           Potassium sulfate: $K^{+1}{}_2(SO_4)^{-2}{}_1$

Step 2. List the elements present in the compound.

K:

S:

O:

Step 3. Find the number of moles present of each element and multiply by the atomic mass values for each element present.

K: 2 x 39 grams = 78 grams

S: 1 x 32 grams = 32 grams

O: 4 x 16 grams = 64 grams

Step 4. Add up the grams.
K: 2 moles x 39 grams/mole = 78 grams
S: 1 mole x 32 grams/mole = 32 grams
O: 4 moles x 16 grams/mole = <u>64 grams</u>
         Formula weight = 174 grams

From these calculations, one mole of potassium sulfate would have the mass of 174 grams.

Example 3. Find the formula weight for one mole of chromium (III) nitrate.

Step 1. Write the formula for the stable compound.
$$\text{Chromium: } Cr^{+3} \qquad \text{Nitrate: } (NO_3)^{-1}$$
$$\text{Chromium nitrate: } Cr^{+3}_{\ 1}(NO_3)^{-1}_{\ 3}$$

Note that there are <u>three</u> nitrate ions in this compound. You can envision the compound like this:

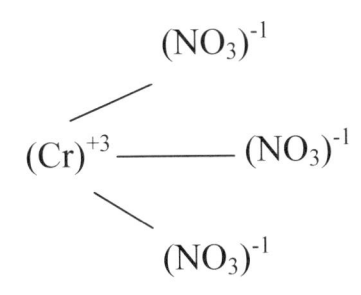

$$(NO_3)^{-1}$$
$$(Cr)^{+3} \longrightarrow (NO_3)^{-1}$$
$$(NO_3)^{-1}$$

$$Cr^{+3}_{\ 1}(NO_3)^{-1}_{\ 3}$$

Steps 2-4.
Cr: 1 mole x 52 grams/mole =   52 grams
N: 3 moles x 14 grams/mole =   42 grams
O: 9 moles x 16 grams/mole =   144 grams (three sets of three oxygens)

Formula weight = 238 grams

Practice finding formula weights by playing Compound Intensity or completing the following practice pages.

Name_____ Date_____

Friendly Chemistry

Lesson 17: Finding Formula Weights Practice –1

Find the formula weights for the following compounds.

1. Sodium oxide

   Formula weight: _____

2. Barium fluoride

   Formula weight: _____

3. Aluminum sulfide

   Formula weight: _____

4. Nickel carbonate

   Formula weight: _____

5. Calcium chlorate

   Formula weight: _____

6. Barium permanganate

Formula weight: _____

7. Zinc hydroxide

Formula weight: _____

8. Potassium nitrate

Formula weight: _____

9. Sodium phosphate

Formula weight: _____

10. Lithium sulfite

Formula weight: _____

11. Calcium acetate

Formula weight: _____

12. Chromium perchlorate

Formula weight: _____

Friendly Chemistry

Name_____ Date_____
Friendly Chemistry

Lesson 17: Formula Weight Practice –2

1. Joel needed to prepare a mole of potassium oxide. How many grams should be get from the stock container?

2. Mary was asked to prepare 3 moles of sodium chloride for a lab activity. How many grams should she get?

3. Frank had 3 moles of lead (II) oxide. How many grams of this compound did he have?

4. Sarah had 10 moles of hydrogen phosphate. How many grams of this compound did she have?

5. Horace had 12 moles of carbon. Julie had 144 grams of carbon. Who had the most carbon?

6. Francis was conducting a lab exercise which required she use 9 moles of iron (III) sulfate. She looked in the chemical closet and found she had a full container of the compound. The label said there was 500 grams in the bottle. Did she have enough? Show your work as proof of your answer.

7. Joey was given 1000 grams of calcium carbonate (limestone). How many moles did he have?

8. Terry had two containers of copper (I) sulfite. One said it had 3.5 moles and the second container's label said it contained 4.75 moles of the compound. How many grams of copper (I) sulfite did Terry have?

Lesson 18: Finding the Percent Composition of Compounds

In our last lesson we learned how to find the formula weight of a compound or how many grams it takes to make one mole of a compound. Let's look at another situation. Suppose that you were asked to prepare a certain amount of a compound, but the amount you were asked to prepare was not a whole number of moles. The amount might be given in parts of a mole or even in grams.

For example, let's pretend that you have been asked to prepare a solution of silver nitrate. For the amount that you need to prepare, you will need 500 grams of the silver nitrate crystals. In order to calculate how much of each ingredient (the silver, nitrogen and oxygen) you would need to prepare 500 grams, it would be helpful to know the percentage each ingredient made of the total 500 grams. The process of finding the percentage of each ingredient is called **finding percent composition**. In other words, you will be finding what part or portion each ingredient contributes to the whole amount (the percentage of the whole compound present).

The first step in finding **percent composition** is to find the formula weight of the compound that you are to prepare. Let's look at an example.

Suppose you would like to prepare 500 grams of sodium chloride. To determine how much sodium and chloride you might need, you will first need to calculate the formula weight of the compound sodium chloride.

$$Na^{+1}_{1}Cl^{-1}_{1}$$

Na: 1 mole x 23 grams/mole = 23 grams
Cl: 1 mole x 35 grams/mole = 35 grams
Formula weight = 58 grams

From these calculations, we see that one mole of sodium chloride has a mass of 58 grams. To determine the percentage of sodium and chloride present, we must compare each ingredient to the formula weight. Let's continue with our example.

Let's start with the sodium percentage. Recall that when finding the percentage of a quantity, one compares the portion they are interested in to the whole quantity. In other words, one compares the "part" to the "whole." In this case, the "part" is the grams of each element present and the "whole" is the formula weight of the compound.

% Sodium = (grams of sodium in one mole / formula weight) x 100%
= (23 grams / 58 grams) x 100%
= 0.3934 x 100%
= 39.6 %

From these calculations we can say that 39.6% of the 500 grams (or any amount, for that matter) of sodium chloride will be made up of sodium or 39.6% x 500 g = 198 grams.

Let's look at the chloride ions now.

% Chloride = grams of chloride in one mole / formula weight x 100%
= 35 grams / 58 grams x 100%
= .6066 x 100%
= 60.4%

From these calculations we can say that 60.4% of the 500 grams (or any amount) of sodium chloride will consist of the chloride ions or 60.4% x 500 grams = 302 grams.

To check your work, add up the two percentages. They should equal 100% or very close to 100% (some small percentages may be lost due to rounding).

$$39.6\% \text{ Na}$$
$$+ \ 60.4\% \text{ Cl}$$
$$100.00\ \%$$

Let's take a look at another example. Suppose you had to prepare 700 grams of barium hydroxide. You will need to know how much of each ingredient will be required. Begin by finding the formula weight for barium hydroxide.

Barium hydroxide

$$Ba^{+2}{}_1(OH)^{-1}{}_2$$

Barium: 1 x 137 grams/mole = 137 grams
Oxygen: 2 x 16 grams/mole = 32 grams
Hydrogen: 2 x 1 grams/mole = 2 grams
Formula weight = 171 grams

Continue by finding the percent composition of barium hydroxide.

% Barium = 137 grams / 171 grams x 100%
= 0.801 x 100%
= 80.1%

% Oxygen = 32 grams / 171 grams x 100%
= 0.186 x 100%
= 18.6%

% Hydrogen = 2 grams / 171 grams x 100%
= 0.012 x 100%
= 1.2%

To check your work, add up the percentages of each ingredient.

$$80.1\% \text{ (Ba)} + 18.6\% \text{ (O)} + 1.2\% \text{ (H)} = 99.9\%$$

(Note that a small percentage was lost due to rounding as we made our calculations.)

To complete this problem, take the percentage of each ingredient (barium, oxygen and hydrogen) of the total amount of the barium hydroxide desired.

Ba: 80.1% of 700 grams = 560.7 grams
O: 18.6% x 700 grams = 130.2 grams
H: 1.2% of 700 grams = 8.4 grams

We can say, therefore, that to make 700 grams of barium hydroxide, you must acquire 560.7 grams of barium, 130.2 grams of oxygen and 8.4 grams of hydrogen.

To summarize this lesson, we found to find the percent composition of a compound, the first step was to write the correct formula for the compound. The next step was to find the formula weight of the compound. Finally, the number of grams from each constituent making up the compound is divided by the total formula weight. To check the calculations, the sum of the percentages is found.

Practice these skills by completing Lesson 18 Practice Pages and then by playing Compound Intensity (level 2).

Friendly Chemistry

Name_____ Date_____

Friendly Chemistry

Lesson 18
Finding Percent Composition of Compounds –1

Find the percent composition of each compound below.

1. hydrogen chloride

   Formula weight: _____ Percent Composition: _____

2. sodium carbonate

   Formula weight: _____ Percent Composition: _____

3. magnesium sulfate

   Formula weight: _____ Percent Composition: _____

4. chromium nitrate

   Formula weight: _____ Percent Composition: _____

5. zinc acetate

   Formula weight: _____ Percent Composition: _____

S161

6. magnesium chlorate

Formula weight: _____ Percent Composition: _____

7. potassium nitrite

Formula weight: _____ Percent Composition: _____

8. aluminum phosphate

Formula weight: _____ Percent Composition: _____

9. calcium chloride

Formula weight: _____ Percent Composition: _____

10. ammonium chromate

Formula weight: _____ Percent Composition: _____

Friendly Chemistry

Name_____ Date_____

Friendly Chemistry

Finding Percent Composition of Compounds –2

Find the percent composition of each compound below.

1. Magnesium hydroxide

   Formula weight: _____ Percent Composition: _____

2. Barium acetate

   Formula weight: _____ Percent Composition: _____

3. Magnesium sulfite

   Formula weight: _____ Percent Composition: _____

4. Aluminum nitrite

   Formula weight: _____ Percent Composition: _____

5. Zinc carbonate

   Formula weight: _____ Percent Composition: _____

6. Calcium hypochlorite

Formula weight: _____ Percent Composition: _____

7. Hydrogen fluoride

Formula weight: _____ Percent Composition: _____

8. Sodium phosphate

Formula weight: _____ Percent Composition: _____

9. Lithium chloride

Formula weight: _____ Percent Composition: _____

10. Ammonium cyanide

Formula weight: _____ Percent Composition: _____

Lesson 19: Writing Empirical Formulas

In the last lesson, we completed a large amount of compound formula writing as well as calculation of percentages and amounts of ingredients required to prepare various chemical compounds. In this lesson, we will investigate the procedures followed in taking a report or analysis of an unknown compound and writing the chemical formula that we suspect that unknown chemical compound to be. In other words, we will take clues about an unknown chemical and figure out its identity based upon those clues.

There is an entire field of study known as **quantitative analysis** where students and chemists learn and develop new ways to analyze unknown chemical compounds that have been produced in the laboratory. Most of these unknown chemical compounds are products and byproducts of chemical reactions intended to create or produce a specific product. For example, in the process of developing an improved fabric coating that could repel water from all-weather jackets while allowing body moisture to evaporate, chemists may produce several potential products and byproducts along the way. Each of these products must be analyzed for its components so that those people using the product understand it properties and capabilities. In addition, all of the byproducts must be analyzed for their components to insure that they are handled and disposed of, if necessary, in the proper ways.

In this lesson we take a previously prepared analysis of an unknown chemical compound and with some relatively simple calculations, arrive at a chemical formula for that unknown compound. In the title of this chapter, the term **empirical formula** is used. In this context, empirical formula means a formula indicating the most reduced whole-number ratio of ingredients found in the compound. Look at these chemical formulas for sodium chloride below:

$$Na^{+1}_{350}Cl^{-1}_{350} \quad \text{Ratio } (350:350)$$

$$Na^{+1}_{200}Cl^{-1}_{200} \quad \text{Ratio } (200:200)$$

$$Na^{+1}_{78}Cl^{-1}_{78} \quad \text{Ratio } (78:78)$$

$$Na^{+1}_{1}Cl^{-1}_{1} \quad \text{Ratio } (1:1)$$

Although all of these chemical formulas represent the very same compound (sodium chloride), the ratio of sodium ions to chloride ion in the *last* formula is in the most reduced form. **It is the chemical formula in which the components are written in the most reduced, whole number ratio, that makes up an empirical formula for a compound.**

You might ask, "How do I know when I've reached the empirical formula for a particular compound?" The answer is that when you can no longer reduce the ratio (or fraction relationship) between the two or more constituents within the compound using only whole numbers (1, 2, 3, etc.), you have reached the empirical formula.

Below are some chemical formulas for some compounds. Circle the ones that are empirical formulas.

H_2O_1 H_4O_2 $K_3(PO_4)$ $K_6(PO_4)_2$ $Cr_1(NO_2)_3$

Note that all of the chemical formulas we have presented in previous chapters have been written as empirical formulas. Now that you can recognize an empirical formula, let's continue with how to go about finding and writing the empirical formula of a chemical compound based upon an analysis of an unknown compound you have received. The best way to learn this procedure is to follow an example.

Suppose you have received the following analysis for an unknown chemical compound you have just produced.

> The analysis for your compound is:
> 20% calcium
> 80% bromine

Based upon the information received in the analysis, you can tell that you have a compound which contains calcium and bromine. To find the correct empirical formula for this compound you must first find the number of **moles** of each ingredient present and then reduce that ratio of moles to the nearest whole number ratio.

To begin finding the empirical formula for this unknown chemical compound, **assume that you have 100 grams of the compound**. (Assuming you have 100 grams of the compound makes the math calculation somewhat easier!) Based upon the analysis, you might then say:

20% of the 100 gram sample represents the calcium ingredient, which is the same as saying . . .
20% x 100 grams = grams of calcium present
0.20 x 100 grams = grams of calcium present
20 grams = grams of calcium present.

And for the bromine ingredient, we could say:
80% of the 100 gram sample represents the bromine ingredient, which is the same as saying . . .
80% x 100 grams = grams of bromine present
0.80 x 100 grams = grams of bromine present
80 grams = grams of bromine present.

Therefore, in our 100 gram sample, we have 20 grams of calcium and 80 grams of bromine. Recall that our goal is find the number of **moles** present of each ingredient. We now have the number of *grams* present of each ingredient, so we need to convert those *grams* of ingredients into **moles** of ingredients.

Recall from the last chapter that the atomic mass value for each element, found on the periodic table of elements, tells the number of grams required to make one mole of that particular element. For example, the atomic mass of carbon is 12 grams. This means that one mole of carbon would have a mass of 12 grams or, yet another way to say it is, that it would require 12 grams of carbon to make one mole of carbon. Therefore, to convert grams to moles we just need to compare the number of grams we have in our sample to the number of grams it takes to make one mole of that element. Let's return to our example of calcium and bromine.

In regard to calcium, we have 20 grams present in our sample. By referring to the periodic table of elements, we can see that calcium has an atomic mass of approximately 40 grams. Based upon this information we can build a ratio which says:

40 grams of Ca : 1 mole of Ca   as   20 grams of Ca : (x) moles of Ca

or

$$\frac{40 \text{ grams of Ca}}{1 \text{ mole of Ca}} = \frac{20 \text{ grams of Ca}}{(x) \text{ moles of Ca}}$$

Since these ratios are considered equal, we can solve for (x) by cross-multiplying.

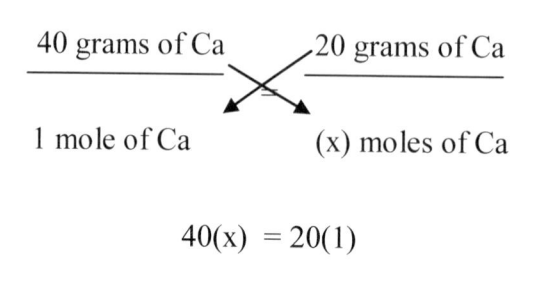

$$\frac{40 \text{ grams of Ca}}{1 \text{ mole of Ca}} \times \frac{20 \text{ grams of Ca}}{(x) \text{ moles of Ca}}$$

$$40(x) = 20(1)$$

$$40(x) = 20$$

Then to solve for (x), we divide both sides by 40.

$$\frac{40(x)}{40} = \frac{20}{40}$$

$$x = 0.5$$

Based upon these calculations, we can say that in our 100 gram sample of unknown compound, we have 20 grams of calcium which represents **0.5 moles of calcium**. In our definition of empirical formula, we stated that the empirical formula must be composed of ingredients in a whole number ratio. Obviously, the moles of calcium we just derived is not a whole number (0.5 is not a whole number). We will not stop at this point to adjust the moles of calcium value to a whole number. Instead, let us determine the number of moles of bromine present. After we have calculated that amount, we will then adjust *both* values to a whole number ratio.

To find the number of moles of bromine present, we will follow the same procedure as we did with finding the moles of calcium present. Find bromine on the periodic table of elements. Note that the atomic mass of bromine is approximately 80 grams. This means that one mole of bromine will have a mass of approximately 80 grams.

From our earlier calculations, we found that in our 100 gram sample of the unknown compound, we had 80 grams of bromine present (80% of the sample was bromine). It may appear obvious that we, therefore, have *one mole* of bromine present in our unknown compound. For completeness of our example, let's examine the calculations to arrive at that value.

$$80 \text{ grams of Br} : 1 \text{ mole of Br} \quad \text{as} \quad 80 \text{ grams of Br} : (x) \text{moles of Br}$$

or

$$\frac{80 \text{ grams Br}}{1 \text{ mole Br}} = \frac{80 \text{ grams Br}}{(x) \text{moles Br}}$$

Since the ratios are considered equal, we can solve for (x)moles of bromine.

$$\frac{80 \text{ grams Br}}{1 \text{ mole Br}} \diagup\!\!\!\diagdown \frac{80 \text{ grams Br}}{(x) \text{moles Br}}$$

$$80(x) = 80$$

$$\frac{80(x)}{80} = \frac{80}{80}$$

$$(x) = 1 \text{ mole}$$

Based upon these calculations, we can indeed say that in our 100 grams sample of unknown compound, we have 80 grams of bromine (80% of 100 grams) which represents **1 mole of bromine**.

Coupled with the number of moles of calcium we calculated earlier, we can say that we have a compound made of **0.5 moles of calcium** and **1 mole of bromine**.

$$Ca_{0.5}Br_1$$

However, this ratio (0.5 : 1) is not a ratio of whole numbers! To convert this ratio to a ratio of whole numbers, simply divide both numbers by the smaller of the two. In this example, the 0.5 value is smaller than the 1 value. Divide both values by the 0.5 value.

$$\frac{Ca_{0.5}Br_1}{0.5 \quad 0.5} = Ca_1Br_2$$

Therefore, our empirical formula for our unknown compound is **Ca$_1$Br$_2$**.

As a simple check of our work, we can examine our empirical formula to see if it is indeed a compound where the charges, when added, equal zero. Recall that calcium (a calcium family member) has two lone valence electrons that it would like to "get rid of" which would result in a net +2 charge while bromine (a halogen family member) has the "desire" to gain one electron to acquire the electron configuration of its noble gas neighbor, krypton. Bromine, therefore, has a net charge of -1.

$$Ca^{+2}_1 Br^{-1}_2 = 0$$

In this case, the charges do equal zero. We can assume that our work was done correctly and the unknown compound has been identified as calcium bromide (Ca$_1$Br$_2$).

Before we move on to another example of how to find the empirical formula of an unknown compound, let's summarize the steps we followed:

1. Examine the analysis provided to determine the elements present in the tested sample.
2. Assume you have been given 100 grams of the unknown sample.
3. Find the number of grams present of each ingredient (based upon the percentages found in the analysis).
4. Convert the number of grams present of each ingredient into moles of each ingredient.

5. Adjust the number of moles present of each ingredient to a whole number ratio by dividing by the smallest value present.
6. Write the empirical formula and check your work by adding up the charges to see that they equal zero.

You may have noticed in the previous example that we rounded the atomic mass values found on the periodic table of elements to whole numbers as we used them in the calculations (40.08 grams/mole of calcium rounded to 40 grams/mole of calcium and 79.909 grams/mole of bromine rounded to 80 grams/mole of bromine). This was done to make the arithmetic easier. As you explore the world of chemistry beyond the scope of this book (especially, the field of quantitative analysis) you will learn when it is appropriate to round values to the nearest whole number and when it is significant not to round values. For the sake of simplifying the concepts and skills required to find the empirical formulas of unknown compounds, we will continue to round the atomic mass values found on the periodic table of elements to whole numbers. In the practice problems that follow the next examples, you can assume that the atomic mass values have been rounded to the nearest whole number.

Let's continue now with another example. Here is an analysis of an unknown compound that you have been asked to identify.

The analysis for your compound is:
11.2% hydrogen
88.8% oxygen

From the analysis, we can see that we have a compound which contains hydrogen and oxygen. Let's begin by assuming we have 100 grams of this unknown compound. In order to find the number of moles present of each ingredient, we will first need to find the number of grams of each ingredient present.

Next we will need to convert *grams* of each ingredient to *moles* of each ingredient. To do so, we will need to find the atomic mass value for hydrogen and oxygen from the periodic table of elements. Hydrogen has an atomic mass of approximately 1 gram/mole and oxygen has an atomic mass of approximately 16 grams/mole.

Hydrogen:

11.2% of our 100g sample = x grams of hydrogen

Which is the same as:

0.112 x 100 g sample = 11.2 grams of hydrogen

Oxygen:

88.8% of our 100 g sample = x grams of hydrogen
Which is the same as:
0.888 x 100 g sample = 88.8 grams of oxygen

Next, we need to convert these *grams* of hydrogen and oxygen into *moles* of hydrogen and oxygen. To do so, we will need to find the atomic mass value for hydrogen and oxygen from your periodic table of elements. Hydrogen has an atomic mass of 1 mole and oxygen has an atomic mass of 16 grams/mole.

So, to convert the hydrogen:

$$\frac{1 \text{ mole of hydrogen}}{1 \text{ gram}} = \frac{x \text{ moles of hydrogen}}{11.2 \text{ grams hydrogen}}$$

By cross multiplying and solving for x we get:

11.2 moles of hydrogen

So far we have $H_{11.2}$.

Now, lets convert our grams of oxygen.

$$\frac{1 \text{ mole of oxygen}}{16 \text{ grams}} = \frac{x \text{ moles of oxygen}}{88.8 \text{ grams}}$$

By cross multiplying and solving for x we get:
5.55 moles of oxygen.

From the calculations above, we can see that we have a ratio of **11.2 moles of hydrogen** for every **5.55 moles of oxygen**. We can write the compound formula as follows.

$$H^{+1}{}_{11.2}O^{-2}{}_{5.55}$$

Our definition of empirical formula states that the ratio must be in whole numbers. To reduce the 11.2 moles : 5.55 moles ratio of ingredients in our compound formula to whole numbers, we divide by the smaller number (5.55 moles).

$$H^{+1}{}_{\frac{11.2}{5.55}}O^{-2}{}_{\frac{5.55}{5.55}}$$

This results in the following compound:

$$H^{+1}{}_{2}O^{-2}{}_{1}$$

Do you recognize this compound? Yes, the unknown compound we were analyzing was *water*! Remember to check the stability of the compound (as well as our calculations) by seeing if the charges add up to zero.

$$H^{+1}{}_{2}O^{-2}{}_{1} = 0$$

In this case, the charges *do* add to zero and we have written an accurate compound formula. This is a good time to make a slight alteration in our formula writing skills. In this example you may not have immediately recognized our unknown formula as being water (most of us recognize water as simply being H_2O). In an effort to simplify the writing of chemical formulas, if only one mole of a cation or anion is present, the subscript can be dropped and it is understood that one mole of that cation or anion is required to make a stable compound. In our example of water, we can write:

$$H^{+1}{}_{2}O^{-2}$$

It is understood that we have two moles of hydrogen cations for every one mole of oxygen anion. We can further simplify the writing of compound formulas by eliminating the writing of the charges which accompany each ion. In this example, we could write the compound formula as:

$$H_2O$$

In our previous example of calcium bromide, we could simplify the writing of the empirical formula by writing:

$$Ca\,Br_2$$

Reviewing the process of finding empirical formulas one more time, we can say that:

1. Assume you've been given 100 grams of the sample.

2. Convert the percentage of each element to grams.

3. Convert the gram amount of each element into moles.

4. Adjust the mole amount of each element to whole numbers by dividing by the lowest mole amount.

Practice finding empirical formulas using the following practice pages.

Name_____ Date_____

Friendly Chemistry

Lesson 19
Finding Empirical Formulas –1

Below are several analyses of unknown chemical compounds. Find the empirical formula for each compound. Then write the name of that compound in the appropriate space.

1. Analysis: 15% fluorine, 85% silver

Empirical formula:_____ Name: _____

2. Analysis: 2.77% hydrogen and 97.3% chlorine

Empirical formula:_____ Name: _____

3. Analysis: 24.58% potassium, 34.81% manganese, 40.5% oxygen

Empirical formula:_____ Name: _____

4. Analysis: 9.66% nitrogen, 87.6% iodine, 2.79% hydrogen

Empirical formula:_____ Name: _____

Wait! There's more on the flip-side!!

5. Analysis: 60.7% chlorine, 39.3% sodium

Empirical formula:_____ Name:_____

6. Analysis: 68% silver, 22% chlorine, 10% oxygen

Empirical formula:_____ Name:_____

7. Analysis: 55% oxygen, 27% phosphorus, 18% lithium

Empirical formula:_____ Name:_____

8. Analysis: 71% bromine, 29% copper

Empirical formula:_____ Name:_____

9. Analysis: 40% oxygen, 45% chlorine, 15% magnesium

Empirical formula:_____ Name:_____

Name_____ Date_____

Friendly Chemistry

Lesson 19
Finding Empirical Formulas –2

Below are several analyses of unknown chemical compounds. Find the empirical formula for each compound. Then write the name of that compound in the appropriate space.

1. Analysis: 39% potassium 14% nitrogen 48% oxygen

Empirical formula:_____ Name: _____

2. Analysis: 32% sodium 23% sulfur 45% oxygen

Empirical formula:_____ Name: _____

3. Analysis: 27% sulfur 33% calcium 40% oxygen

Empirical formula:_____ Name: _____

4. Analysis: 48% oxygen 35% chlorine 17% chromium

Empirical formula:_____ Name: _____

5. Analysis: 59% oxygen 29% sulfur 13% lithium

Empirical formula:_____ Name:_____

6. Analysis: 94% sulfur 6% hydrogen

Empirical formula:_____ Name:_____

7. Analysis: 24% nitrogen 21% carbon 55% zinc

Empirical formula:_____ Name:_____

8. Analysis: 32% nickel 15% nitrogen 52% oxygen

Empirical formula:_____ Name:_____

9. Analysis: 3% hydrogen 50% chromium 47% oxygen

Lab Activity: The Composition of Air

What is air made of? Most likely your first thought is that air is made up of oxygen, because it obvious that we all need the oxygen in the air to survive. But did you know that there are other gases in the air as well? In this lab, you'll take a sample of air and attempt to determine the percentage of air that is made up by oxygen. Then, you'll do a little research to find out, what gases make up the remaining percentage. Let's begin.

Your first problem is how can you separate gases in the air when you can't see what you're trying to separate? Any ideas? Write them here:

If you didn't come up with any ideas, try this approach: use a chemical reaction which can use up the oxygen in your sample of air. Any ideas yet?

Yes, burning or combustion consumes oxygen from the air. What happens when you take a burning candle and place a glass jar over it? If you haven't ever tried that, do so now. Ask your teacher for supplies.

So, what happens?

Back to our question of what percentage of air is made up by oxygen. Knowing that a burning candle will consume oxygen, can you think of a way we might be able to measure how much oxygen the candle might consume?

Ask your teacher for a recommended method for doing this. He or she has a method in mind that works pretty well. After you get things set up, record your data on the table on the next page. If you have another means of accomplishing this task, discuss it with your teacher. If your idea is safe, he or she may allow you to try it out!

Record your data here:

Trial Number	Milliliters in Empty Container (full of air sample)	Amount of water that entered container after candle goes out	Milliliters of oxygen gas consumed by reaction.	Percentage of oxygen in sample.
			Average Percentage:	

Based upon your data, what do you think the average percentage of oxygen is in a sample of air? _____

Now, go do some research to find out what the average amount of oxygen is in the air. Write that value here: _____

Were you very close to this accepted average? _____

What are some reasons why you might have a different result?

Now that you know the percentage of oxygen made up by air, what gases make up the other _____ percent? Write those gases below:

S180

Friendly Chemistry

Name_____ Date_____

Chromatography Lab

DATA TABLE FOR INK SPECIMENS

SOLVENT USED:_____

Ink Sample (make mark here)	Description of Marker or Pen Denote washable or non-washable by using W or NW.	Results - tell what happened as the solvent encountered the spot. List the colors that appeared.

S181

Friendly Chemistry

Name_____ Date_____

Chromatography Lab

DATA TABLE FOR PLANT SPECIMENS

SOLVENT USED:_____

Name of Plant Used	Description of Plant Part — Describe the colors present in this plant part.	Results - tell what happened as the solvent encountered the spot. List the colors that appeared.

S182

Lesson 20: Putting Compounds into Reactions

In the last lesson, we discussed the rationale for learning the skills of writing empirical formulas. We said that in the process of creating new products for consumer use, many byproducts are also produced. We discussed that analysis of these products and byproducts was essential to determine their potential usefulness and possible methods of disposal. In this chapter, we will begin to explore the basic concepts of how chemical compounds interact with each other. We will also discuss how we can record what happens in these interactions in order to share this information with others.

To facilitate our discussion, let's use an analogy. In your life, you have very likely encountered several situations in which you had the opportunity to make new friends. These situations may have been as simple as making a new friend at the merry-go-round on the playground or as exciting as meeting a companion who could "be the one" to share your life adventure. These situations can range from being uneventful to quite eventful. For example, you may make casual conversation with the young man or lady who carries your groceries to the car and never get past phrases like, "It's really a nice day," or "Thanks for not squishing the bread." However, in other situations, your interaction with another person may be much more eventful and a conversation may grow into a lifelong friendship with discussions of very meaningful topics like goals, dreams and aspirations. The range of interactions between chemical compounds is just as wide! The interaction varies from being almost incomprehensible to the very intense and even explosive! Some interactions may be very slow in developing and some may be instant (akin to "love-at-first-sight)!

Let's begin by introducing some terms which are used by chemists to describe a chemical interaction. First of all, the interaction between two or more chemicals, whether they be single elements or compounds, is known as a **chemical reaction**. The participants in a chemical reaction (that is, those chemicals that have found themselves in a situation in which they can interact with each other) are known as the **reactants**. The chemical nature (i.e. reactivity) of these reactants determines whether a reaction will take place. What results from the reaction are known as the **products** of the reaction.

Recall from our earlier discussions how the arrangement of electrons in the atoms' electron clouds determines the reactivity of those atoms. Recall that all atoms have a tendency to seek the stability of a noble gas. By taking the opportunity to react with another chemical, whether it be a **free element** or compound, a reactant might have a chance at acquiring a more stable configuration.

You might compare this idea to that of two talented musicians: one very gifted in playing the bass guitar and the other quite talented in playing the lead guitar. By themselves they may be limited in their performance, but together they can produce quite an outstanding presentation and possibly be more satisfied with their performance. Likewise, reactants interact with each other in an attempt to acquire a more stable, "satisfying" electron configuration.

Now, back to our terms. We stated that the reactants were those chemicals that were exposed to each other in an effort to lead to a reaction. The results of this reaction are known as the **products** of the reaction. The products are those chemicals, whether they be free elements or compounds, which result after two or more reactants interact with each other.

In order to say that a chemical reaction has taken place (and new products made), there must be evidence that a **chemical change** has taken place in the reactants. In our analogy, we might say that a reaction has taken place when our thoughts or ideas change as a result of interacting with our new acquaintance. The evidence of a chemical change varies according to the reactants. The results that are generally considered evident of a chemical change are: (1) a change in the chemical properties of the reactants; (2) a change in the solubility (how well a substance dissolves in another substance) of the reactants; or (3) a change in the color of the reactants.

An example of a change in the chemical properties of the reactants, which is evidence of a chemical change, is how the properties of the free element sodium change drastically when it reacts to form sodium chloride, commonly known as table salt. (We use the term "free" to indicate that the sodium is not bonded to any other element or anion: it is by itself.) Recall that sodium (atomic number 11) has one lone electron in its outermost energy level and will readily lose that single electron to acquire the stability of its noble gas neighbor, neon (atomic number 10). Because of sodium's very strong desire to lose that single electron, it is an extremely reactive element. In fact, it is so reactive that it must be stored under a petroleum product to keep it from reacting with water vapor in the air! These explosive properties are radically changed when the sodium cations form an ionic bond with chloride anions to make sodium chloride. Recall that chlorine is a member of the halogen family and has the desire to accept one electron to achieve the stability of the noble gas family member, argon. Sodium chloride, as you well know, is a very stable compound. You can heat table salt and nothing happens. You can mix it with almost anything and little happens. Most people can eat table salt and nothing happens (when it is eaten in moderate quantities). This drastic change in chemical properties (sodium's change from high reactivity to low reactivity) is very apparent evidence that a chemical change in the reactants has occurred.

The presence of a color change as evidence of a chemical change can readily be observed when jewelry (often made of copper) turns your skin a greenish-blue color. The reaction between the substances on your skin and the copper jewelry causes the copper to change color. Another example is the whitening characteristics of sodium hypochlorite, commonly known as chlorine bleach. You may have had the unfortunate experience of having bleach splash onto your favorite shirt or jeans. The resulting white spots are evidence that a chemical change has occurred. Scientists who use hydrogen peroxide to whiten bones of an animal skeleton before they put it back together, witness a chemical change, also. The simple reaction of a piece of metal rusting when exposed to moisture (color change of silvery-gray to reddish brown) is yet another example of a chemical change.

Detecting a change in solubility as evidence of a chemical change may not be as commonplace around the house as detecting changes in chemical properties or color changes. You may have experienced the chemical change which occurs when you mix vinegar with baking soda. The reactants here are acetic acid (vinegar) and sodium bicarbonate (baking soda). The bubbles which result are filled with carbon dioxide gas which is no longer soluble in the vinegar and rises to the surface. This change in solubility is evidence that a chemical change has taken place.

Another example of a change in solubility (which is evidence that a chemical change has occurred) is in the preparation of cheese. As vinegar is slowly added to warmed milk, the milk proteins clump or coagulate and no longer are evenly dispersed in the milk. These clumped milk proteins form the cheese and the remaining liquid forms the whey. In the chemistry laboratory, this formation of an insoluble product is known as **precipitation**. The insoluble product is known as a **precipitate**. Creating a precipitate is one way chemists can separate chemical compounds.

Now, not all reactants immediately begin reacting once they are exposed to each other. Some reactants, like shy individuals, may need to be "introduced" to each other and some energy applied to the situation in order to "get things rolling." This needed energy may be as simple as a gentle shake or stir or as complicated as application of external heat or an electric spark! For example, when baking a cake, the recipe requires that the ingredients are vigorously stirred or beaten. This is done to insure that the baking powder gets exposed (or "introduced") to the other ingredients which results in the formation of carbon dioxide bubbles which causes the cake to rise. Another example is the heating that is required to make candies such as peanut brittle. In order for the sugars in the candy to chemically change, heat must be applied.

Other examples are photoflash bulbs. A tiny spark of electricity is applied to a tiny aluminum wire inside the oxygen-filled bulb. The aluminum wire is ignited and reacts vigorously with the oxygen to emit an intense flash of light. Without the spark to get the reaction started, nothing will happen between the aluminum wire and the oxygen.

In other words, energy may have to be applied to a set of reactants in order for a reaction to get underway. Some reactions require that energy be continuously applied for the reaction to continue, whereas others may only need the initial energy to get things started. An example is the simple kitchen match. The heat generated through the friction of striking the match is all that is needed to get the phosphorus on the tip of the match to react with oxygen in the air. Once the reaction is started, the remaining chemicals in the match begin and continue to react, including the wood which undergoes obvious chemical changes (new properties and color changes!).

Reactions which require an input of energy in order for the reaction to get underway are known as **endothermic reactions**. The prefix endo- means "putting in" and the root word "thermic" means "heat" (as in thermos, thermostat or thermal blanket). You can see that the term endothermic means that heat or energy must be added to the reaction for it to commence. The energy that is "put into" the reaction may be stirring, shaking, heating or applying electrical or light energy to the reactants.

On the other hand, some reactions produce or release energy as the reaction proceeds. The energy that is released may be in the form of heat, light or sound. These reactions in which energy is produced or released are known as **exothermic reactions**. The prefix exo- means "leaving or exiting" and, again, the root word "thermic" means heat or energy. Together, we have the word exothermic, which indicates that energy is leaving the reaction.

There are many examples of exothermic reactions . The process of burning or combustion is an exothermic reaction. The process of striking a match may be an endothermic reaction, but once the reaction has begun, the remaining reactions are exothermic. Energy in the form of heat and light from the flame of the burning match are evidence of an exothermic reaction. Burning a candle and the filament in a light bulb are also examples of exothermic reactions. The gasoline and oxygen reaction in an automobile engine is quite an exothermic reaction. A tremendous amount of heat and sound energy are released. In the process of making lye soap, the lye crystals (sodium hydroxide) must be dissolved in water. The lye reacts with the water producing a large amount of heat. The soap-maker must be aware of this heat production and allow the mixture to cool before mixing it with the fat or tallow to make the soap. In addition, the use of lye or strong acids, such as sulfuric acid, to open clogged drains can be an exothermic reaction and all precautions on the drain opener label should be followed closely!

Below are some descriptions of chemical reactions that you may not be familiar with that can be performed in a laboratory. See if you can identify them as being endothermic or exothermic reactions or, possibly both! Correct answers follow the reactions.

1. When hydrochloric acid (HCl) and water (H_2O) are placed into the same container, the container becomes very warm and sometimes even hot to the touch. Is this reaction considered endothermic, exothermic or both?

2. When magnesium ribbon (a form of the element magnesium which has been flat-

tened into a narrow, ribbon-like strip) is heated with a match, the magnesium begins reacting with oxygen in the air to give off light so intense that it can damage your eyes if observed directly. In addition, intense heat can also be felt near the reacting magnesium. Is this reaction considered endothermic, exothermic or both?

3. An interesting reaction takes place when barium hydroxide and ammonium chloride are stirred together. As reactants, they are both white crystals, but as they are vigorously stirred together they become a colorless liquid. If the beaker or container in which the reaction is taking place is placed over a puddle of water on a wooden block, the water quickly freezes causing the block to become adhered to the bottom of the beaker or container. Is this reaction considered endothermic, exothermic or both?

Answers:

Reaction 1: Exothermic

Reaction 2: Both; endothermic when heated with match and exothermic when light is produced.

Reaction 3: Endothermic; both when mixing to get reaction underway and then following as water freezes beneath the container.

When making new acquaintances and friends, we, like some reactants, may be permanently changed into a new "product." On the other hand, we may be only temporarily altered and soon return to our previous state. We can say that some chemical reactions are reversible and some are irreversible. To indicate when a reaction may reverse or is unidirectional (irreversible), we can utilize a system of notation to represent chemical reactions.

To describe what is happening in a chemical reaction we can write the reaction using words or symbols. The written description of a chemical reaction (whether in words or symbols) is known as a **chemical equation**. Let's look at some examples of chemical equations.

When hydrogen and oxygen are placed under appropriate conditions water can be produced. To write this reaction in words, we might say:

Hydrogen and oxygen react to produce water.

We can write the same chemical reaction using symbols. The chemical formulas for each reactant are written using an addition sign (+) to indicate that the chemicals are being combined. To indicate that a chemical reaction has taken place, an arrow is written (following the reactants) in the direction of the products which are written to the right of the arrow. Let's look at the hydrogen and oxygen reaction written, now using symbols:

$$H_2 + O_2 \rightarrow H_2O$$
$$\text{Reactants} \qquad \text{Products}$$

Note that the reactants are written to the left of the arrow and the products are written to the right of the arrow. To indicate if the reaction is irreversible, a single arrow is used. If the reaction is reversible, two arrows are used: one in each direction. Let's look at an example.

Hydrogen and iodine react to produce hydrogen iodide (the reaction is reversible)

$$H_2 + I_2 \rightleftarrows HI$$

In addition to writing the reactants and products, we can also use symbols to indicate the phase or state that each component is in during the reaction. The three phases or states of matter we are referring to are solids, liquids or gases. To indicate these states of matter in the equation, the symbol (s) is used to indicate a solid, the symbol (aq) or (l) is used to indicate a liquid or aqueous form, and the symbol (g) to indicate a gaseous form. Look at this example.

Gaseous nitrogen dioxide and water react to produce liquid nitric acid and gaseous nitric oxide.

$$NO_{2(g)} + H_2O_{(l)} \rightarrow HNO_{3(l)} + NO_{(g)}$$

If certain conditions are required for the reaction to take place, those conditions can be written over the arrow in the chemical equation. For example, the word heat may be written over the arrow or the specific temperature can be written if it is known. Here is an example.

Solid ammonium nitrate reacts at 200 C. to form gaseous nitrous oxide and water.

$$NH_4(NO_3)_{(s)} \xrightarrow{200\ C} N_2O_{(g)} + H_2O_{(l)}$$

Another notation that can be written over or beneath the arrow to indicate specific reaction conditions is the word **catalyst**. A catalyst is a substance which is added to a reaction to increase the speed of the reaction. The catalyst itself, however, is not changed in the reaction. Look at this example.

Sulfur dioxide gas and oxygen gas will react under heat and in the presence of a catalyst to produce sulfur trioxide gas.

$$SO_{2(g)} + O_{2(g)} \xrightarrow[\text{catalyst}]{\text{heat}} SO_{3(g)}$$

To review, in this lesson we learned how chemical compounds interact with one another. The compounds that were put into the reaction are called reactants where the compounds which result from the reaction are called products. We learned that we can add additional information to the written chemical reaction by telling what phase (solid, liquid or gas) the reactants or products may be found in as well as if heat or some other sort of catalyst is necessary for the reaction to take place.

As we continue working with chemical reactions, we will come across certain elements which are gases at room temperature and "like" to exist as two atom pairs. For example, oxygen gas at room temperature likes to exist as O_2 (two atoms of oxygen existing as a pair). Gases, like oxygen gas are called **diatomic gases** where the prefix di– refers to two, therefore we have the term two-atomed gases. In addition to oxygen gas, there are several other diatomic gases. The diatomic gases that you will encounter in this course are:

Oxygen gas……..O_2
Hydrogen gas……….H_2
Chlorine gas………..Cl_2
Bromine gas ………..Br_2
Nitrogen gas ………N_2
And Iodine (while not a gas at room temperature)……..I_2

Practice rewriting word chemical equations into chemical formula equations on the pages which follow. In addition, you will be conducting a number of lab activities to help you see firsthand some chemical reactions!

Friendly Chemistry

Name_____ Date_____
Friendly Chemistry

Lesson 20
Writing Chemical Reactions

Below you will see several chemical reactions. Re-write these word reactions into chemical formula reactions. The first problem has been completed for you!

1. Hydrogen oxide decomposes into hydrogen gas and oxygen gas.

$$H_2O \longrightarrow H_2 + O_2$$

2. Calcium and hydrogen oxide react to yield calcium hydroxide and hydrogen gas.

3. Sodium hydroxide and hydrogen chloride react to yield sodium chloride and water.

4. Ammonium nitrite decomposes to yield water and nitrogen gas.

5. Hydrogen cyanide and sodium sulfate are the products when hydrogen sulfate and sodium cyanide are combined.

6. Zinc oxide and hydrogen chloride react to yield water and zinc chloride.

7. Elemental carbon © and aluminum oxide react to produce aluminum and carbon dioxide (CO_2).

8. Barium carbonate is the result of the reaction between barium oxide and carbon dioxide.

9. Potassium chloride and fluorine gas react to yield potassium fluoride and chlorine gas.

10. Zinc metal and sulfur react to yield zinc sulfide.

11. When calcium is mixed with water, calcium hydroxide and hydrogen gas are produced.

12. Barium carbonate is the product of the reaction between barium oxide and carbon dioxide.

13. Hydrogen sulfate and sodium hydroxide react to yield sodium sulfate and water.

14. Water and nitrogen gas react to yield ammonium nitrite.

15. Iron and copper (II) sulfate react to yield iron (II) sulfate and copper.

16. Ammonium chloride and calcium hydroxide react to produce calcium chloride, ammonia (NH_3), and water.

17. Hydrogen chloride and lead (II) sulfide are the products of the reaction between hydrogen sulfide and lead (II) chloride are combined.

18. Potassium chlorate will decompose quickly into potassium chloride and oxygen gas.

NAME_____ DATE_____
FRIENDLY CHEMISTRY

Krazee Krunch Lab
My Observations

Ingredient	Before Cooking	During Cooking	After Cooking
Popcorn			
Butter or margarine			
Light Corn Syrup			
Sugar			

Lesson 20

NAME_____ DATE_____
FRIENDLY CHEMISTRY

Peanut Brittle Lab
My Observations

Ingredient	Before Cooking	During Cooking	After Cooking
Sugar			
Corn Syrup			
Water			
Salt			
Baking Soda			
Peanuts			
Butter			

NAME_____ DATE_____
FRIENDLY CHEMISTRY

"pH Phun" Lab Data Table

Test Solution	Initial Color	Color after Adding Cabbage Juice	Acid/Base/Neutral?	Additional Observations
1				
2				
3				
4				
5				
6				
7				
8				
9				
10				
11				
12				

Lesson 20

NAME_____ DATE_____
FRIENDLY CHEMISTRY

Baking Soda and Vinegar Balloon Lab
Data Table

Trial #	Amount of Baking Soda	Diameter of Balloon	Other Observations
1			
2			
3			
4			
5			
6			
7			
8			
9			
10			
11			
12			
13			
14			
15			
16			
17			
18			
19			
20			
21			
22			
23			
24			
25			
26			
27			
28			
29			
30			

Graph Paper for Baking Soda Labs

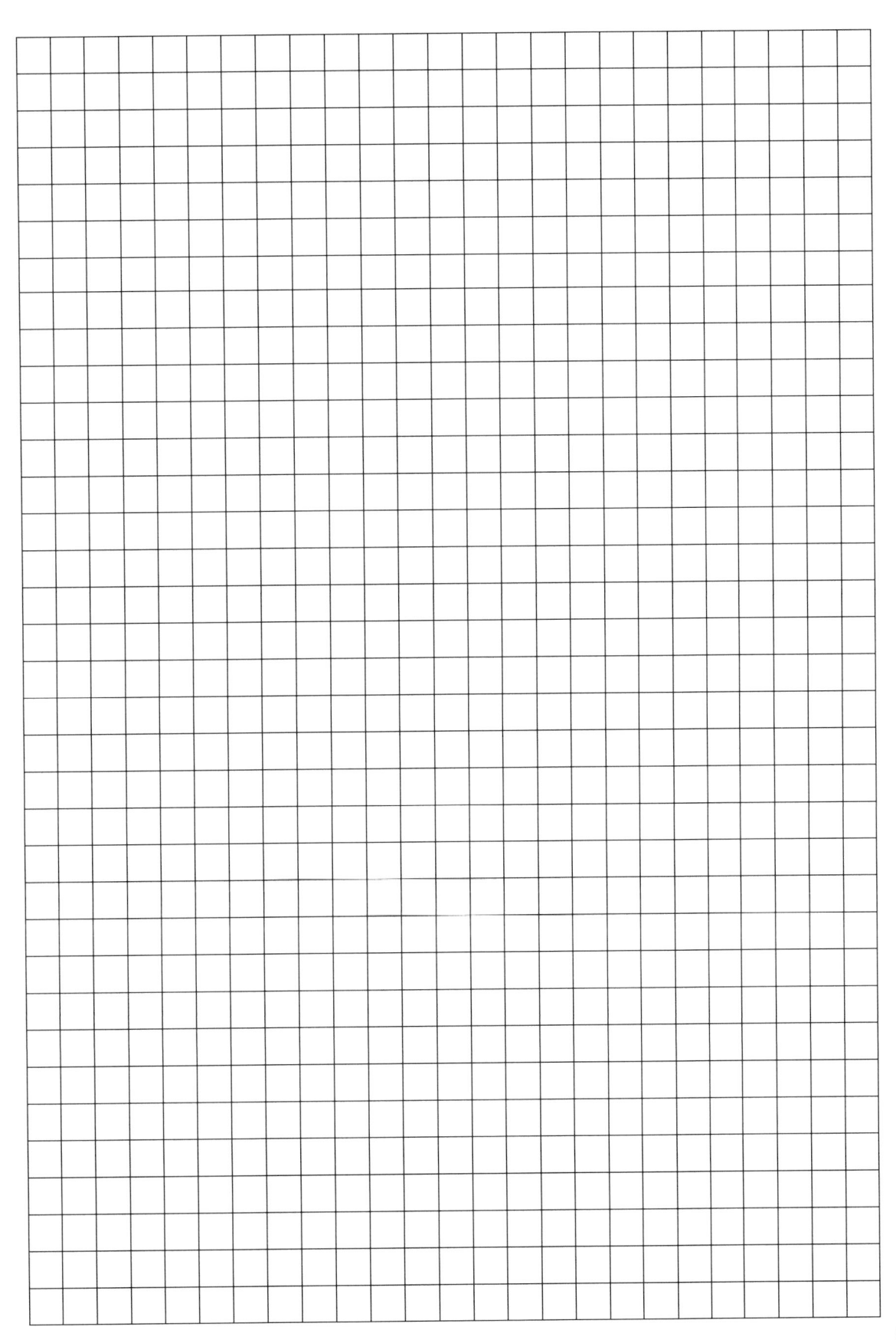

Lesson 20

NAME_____ DATE_____
CHEMISTRY

Endothermic Reactions

Time	Temperature	Observations

TEMPERATURE

TIME

S198

NAME_____ DATE_____
CHEMISTRY

Exothermic Reactions

Time	Temperature	Observations

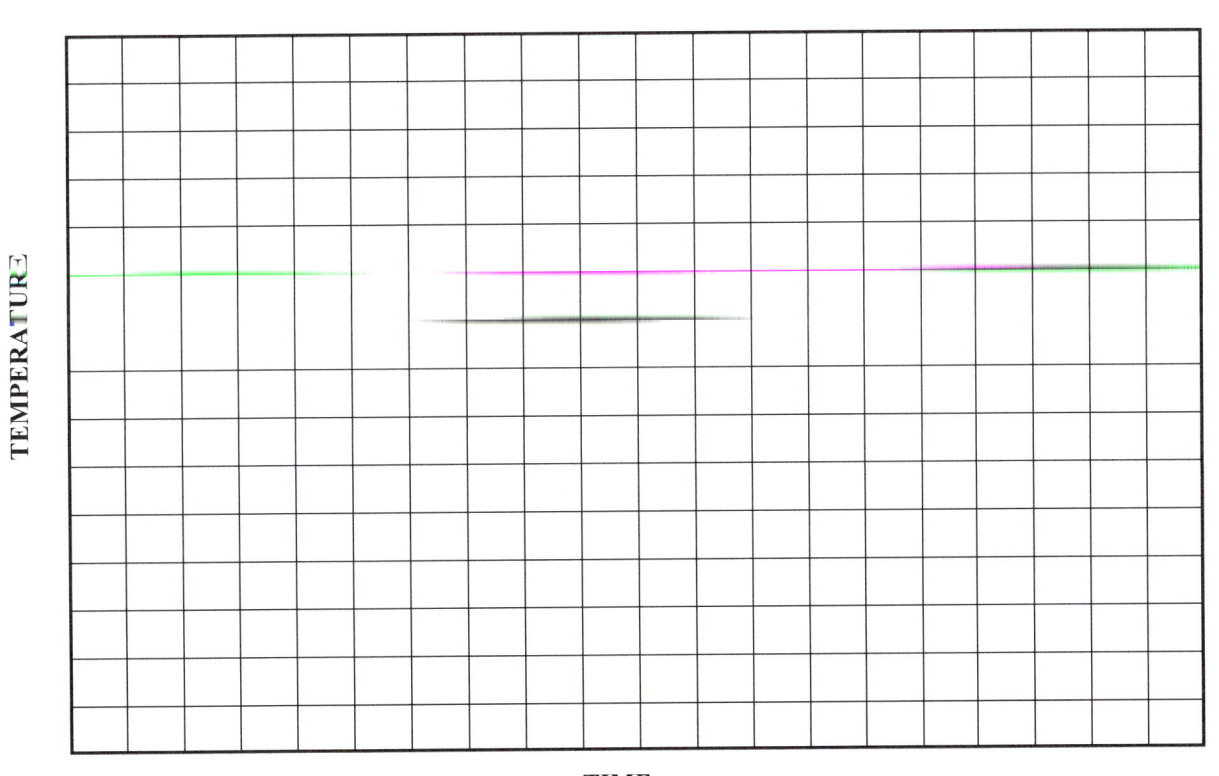

TEMPERATURE

TIME

Lesson 20

Lesson 21: Balancing Chemical Equations

    In addition to allowing us to communicate with others about chemical reactions we have observed, chemical equations are useful in determining relative amounts of reactants which may be necessary to produce a desired amount of product. Chemical equations are also useful in predicting the amount of a product possible from a known amount of ingredient. In order to gain the skills necessary to determine amounts of ingredients necessary or to predict an amount of product, we must first understand a basic principle of chemistry. This basic principle is known as the **Law of Conservation of Matter**. Although this principle is called a law, it is not like a governmental law such as a speed limit. Instead, it is a basic premise which states that matter is neither created nor destroyed in a chemical reaction; it only changes its form. In other words, when two or more reactants interact with each other, there is no new matter created as a product nor is there any matter destroyed in the process. The products are only the reactants in a different form. An easy way to think about this principle is: what-you-put-in-is-what-you-get-out-only-in-a-different-form.

One more point needs to be made. The number of moles of each element of ingredient that you put into a reaction must equal the number of moles of each element of product that you get out of the reaction. You cannot create any new matter or destroy any in the process. You might compare this law to the making of a fruit salad. You may take four apples, four oranges and twenty grapes for your salad, slice them and toss them into a bowl. If you examine the product (the salad) you will see that you still have four apples, four oranges and twenty grapes. They have only changed form. You neither created nor destroyed any matter in the process.

In the reaction of hydrogen gas and oxygen gas to create liquid water, you can see that we put hydrogen atoms and oxygen atoms into the reaction and got hydrogen and oxygen atoms out in the form of the product, water. No additional kinds of atoms were created or destroyed in the process.

$$H_2(g) + O_2(g) \rightarrow H_2O_1(l)$$

If we examine the equation written above, we see that the number of moles of hydrogen atoms on the reactant side of the equation equals the number of moles of hydrogen atoms on the product side (which is 2). However, the number of moles of oxygen atoms on the reactant side (2) does not equal the number of moles of oxygen atoms on the product side (1).

$$H_2(g) + O_2(g) \rightarrow H_2O_1(l)$$

The Law of Conservation of Matter states that matter can neither be created nor destroyed, yet it *appears* here that we have destroyed one mole of oxygen in the process of creating water! Because we cannot destroy matter, we must go through a process known as **balancing our chemical equation** to accurately represent the chemical reaction which has taken place. In order to balance the chemical equation, the number of moles of each ingredient must equal the number of moles of each product. In other words, the amount that goes into the reaction must equal the amount that comes out of the reaction.

Let's go back to our example of the formation of water from hydrogen and oxygen. Here is the original equation (we have removed the notations indicating the phase of each component to simplify this notation):

$$H_2 + O_2 \rightarrow H_2O_1$$

In order to balance a chemical equation, such as this one, we can use multiples of each component. Once each component in the equation has been correctly written (each compound's charges add to zero), we can use numbers known as **coefficients** to indicate a multiple of one or several of the components. These coefficients are always whole numbers and are written just before the component of which a multiple is desired. In our example of the formation of water, we can make the following analysis.

$$H_2 + O_2 \rightarrow H_2O_1$$

Reactants	Products
H: 2	H: 2
O: 2	O: 1

In order to make the number of moles of oxygen on the product side of the equation (1) equal the number of moles of oxygen on the reactant side of the equation (2), we can place a coefficient of 2 in front of the compound of water on the product side.

$$\downarrow$$

$$H_2 + O_2 \rightarrow 2H_2O_1$$

Note that once we place a coefficient in front of a particular compound or element, all constituents of that compound are subject to that coefficient. In other words, each part of that compound is also multiplied by the coefficient. In the case of our example, not only do we multiply the number of moles of oxygen by 2, but we must also multiply the number of moles of hydrogen by two. Here is the equation again with the new analysis of moles present on each side of the equation.

$$H_2 + O_2 \rightarrow \mathbf{2}H_2O_1$$

Reactants	Products
H: 2	H: 4
O: 2 ⟷	O: 2

Note that the number of oxygen atoms are now equal! However, the number of hydrogen atoms has become *not* equal. On the reactant side of the equation there are 2 moles of hydrogen and, on the product side we now have **4** moles of hydrogen. In order to correct this situation, we can apply a coefficient of 2 in front of the hydrogens on the reactant side of the equation.

↓

$$2H_2 + O_2 \rightarrow 2H_2O_1$$

If we now compare the number of moles of each component on the reactant side to the number of moles of each component on the product side of the equation, we can see that we have "balanced" this equation.

$$2H_2 + O_2 \rightarrow 2H_2O_1$$

Reactants	Products
H: 4 ⟷	H: 4
O: 2 ⟷	O: 2

You may be wondering why we didn't just insert the coefficient of 2 in between the hydrogen and oxygen in the water molecule ($H_2 2O_1$). If we did so, it would have given us the proper number of oxygen atoms and would not have affected the number of hydrogen atoms. However, the compound of water does not exist with 2 hydrogen atoms and 4 oxygen atoms as would be indicated if we inserted the 2 between the two compound constituents! We must not fall to the temptation to insert coefficients *within* a compound. The coefficients have to be placed **in front** of the compounds, and all constituents of the compound which immediately succeed that coefficient are subject to it.

Before we look at more examples of balancing chemical equations, let's review the process. After correctly writing the word equation, we next write the equation using chemical formulas and an arrow indicating the direction of the reaction. We learned it is customary to write the reactants on the left side of the arrow and the products on the right side of the equation.

Once each component is correctly written (remember to be sure that the charges on each compound add to zero and that some gases exist as diatomic molecules), we can analyze the number of moles of each component. If the number of moles of any component on the reactant side of the equation does not equal the number of moles on the product side of the equation, we can use coefficients to indicate a multiple of that component. The coefficient is applied to the entire compound which immediately follows that coefficient. **If no coefficient is written, we can assume that one unit of that reaction component is present.** Once the number of moles on each side of the equation is equal, the equation is said to be balanced. Remember that a balanced chemical equation is an accurate representation of the chemical reaction described.

Let's look at another example. By mixing silver nitrate and the free element nickel, we can produce the compound nickel nitrate and the free element silver. Begin by writing the formula equation.

silver nitrate and nickel react to yield nickel nitrate and silver

$$Ag^{+1}_{1}(NO_3)^{-1} + Ni \rightarrow Ni^{+2}_{1}(NO_3)^{-1}_{2} + Ag$$

After checking that each compound is correctly written (charges add up to zero), we can continue by analyzing what is present on each side of the equation. Here is the equation again and the analysis. Note that since we had "nitrates" both on the reactant and product side, we left them "together" (did not separate and count individual nitrogen and oxygens)

$$Ag^{+1}_{1}(NO_3)^{-1} + Ni \rightarrow Ni^{+2}_{1}(NO_3)^{-1}_{2} + Ag$$

Reactants   Products

Ag: 1   Ag: 1

(NO$_3$): 1   (NO$_3$): 2   Not Balanced Here!

Ni: 1   Ni: 1

If we place a coefficient of 2 in front of the silver nitrate on the reactant side of the equation, we can "alleviate" some of the imbalance.

$$\downarrow$$
$$2Ag^{+1}{}_1(NO_3)^{-1} \;+\; Ni \;\rightarrow\; Ni^{+2}{}_1(NO_3)^{-1}{}_2 \;+\; Ag$$

The resulting analysis appears like this.

Reactants	Products
Ag: 2	Ag: 1
(NO$_3$): 2	(NO$_3$): 2
Ni: 1	Ni: 1

By referring back to our analysis, you can see that only the number of moles of silver are not balanced. To balance the moles of silver, we can write a coefficient of 2 in front of the silver on the product side of the equation.

$$\qquad\qquad\qquad\qquad\qquad\qquad\qquad\qquad\qquad\qquad\qquad \downarrow$$
$$2Ag^{+1}{}_1(NO_3)^{-1} \;+\; Ni \;\rightarrow\; Ni^{+2}{}_1(NO_3)^{-1}{}_2 \;+\; 2Ag$$

An analysis now reveals that we have balanced the equation.

$$2Ag^{+1}{}_1(NO_3)^{-1} \;+\; Ni \;\rightarrow\; Ni^{+2}{}_1(NO_3)^{-1}{}_2 \;+\; 2Ag$$

Reactants	Products
Ag: 2	Ag: 2
(NO$_3$): 2	(NO$_3$): 2
Ni: 1	Ni: 1

Let's examine one more example. When water is mixed with free calcium, calcium hydroxide and the explosive hydrogen gas is produced.

calcium and water   react to yield   calcium hydroxide   and   hydrogen gas
$$Ca \;+\; H_2O \;\rightarrow\; Ca^{+2}{}_1(OH)^{-1}{}_2 \;+\; H_2$$

Note that the hydrogen gas that is produced is one of the diatomic gases. (Recall that in addition to hydrogen, the gases oxygen, fluorine, chlorine, bromine, iodine and astatine occur as diatomic molecules i.e. sets of two: O_2, F_2, Cl_2, Br_2, I_2 and At_2).

$$Ca + H_2O \rightarrow Ca^{+2}_{1}(OH)^{-1}_{2} + H_2$$

An initial analysis shows that the equation as written is not balanced.

Reactants	Products
Ca: 1	Ca: 1
H: 2	H: 4
O: 1	O: 2

Note that we collected all of the moles of hydrogen on the product side into one value (2 from the calcium hydroxide and 2 from the hydrogen gas). There appears to be a "problem" with the number of moles of hydrogen and the number of moles of oxygen. By placing a coefficient of 2 in front of the water on the reactant side, we can balance the equation.

$$Ca + 2H_2O \rightarrow Ca^{+2}_{1}(OH)^{-1}_{2} + H_2$$

Reactants	Products
Ca: 1	Ca: 1
H: 4	H: 4
O: 2	O: 2

Note that in this example it was important that we "dismantled" each component in the reaction to correctly balance the equation. Had we left the hydroxide ion as one unit in our analysis, it would have been very difficult to adjust the coefficients to balance the number of moles of hydrogen and oxygen.

In looking back at these three examples of balancing equations, you might think that all equations come "unbalanced." This is not the case as you can see in the following example.

When sodium hydroxide (lye) and hydrogen chloride (hydrochloric acid) are allowed to react, sodium chloride (table salt) and water result.

sodium hydroxide and hydrogen chloride react to yield sodium chloride and water

$$Na^{+1}_{1}(OH)^{-1}_{1} + H^{+1}_{1}Cl^{-1}_{1} \rightarrow Na^{+1}_{1}Cl^{-1}_{1} + H^{+1}_{2}O^{-2}_{1}$$

Reactants	Products
Na: 1	Na: 1
O: 1	O: 1
H: 2	H: 2
Cl: 1	Cl: 1

This reaction "came" balanced!

In attempting to balance some equations, you may find it beneficial to write the compound of water in different ways. In some cases writing water as hydrogen hydroxide, H(OH), makes the job easier should you have compounds with hydroxide in them on the opposite side of the arrow. In other cases, writing water as H_2O works better.

Another question you may have pondered is if the coefficient 2 was the only coefficient used to balance equations. Although 2 is used quite often, coefficients of 3, 4 and sometimes 6 are commonly used. Any whole number can be used as a coefficient.

One final reminder is that once you place a coefficient in front of a compound, it has its effect upon the entire compound. Sometimes it may be tempting to insert a coefficient *within* a compound. Once again, the coefficient must go "out front" of a compound.

Practice balancing chemical equations by completing the following Lesson 21 practice pages.

NAME_____ DATE_____

FRIENDLY CHEMISTRY

Lesson 21
Balancing Chemical Equations –1

Write and balance each chemical equation below. Ask if you need help!

1. hydrogen gas and chlorine gas react to yield hydrogen chloride

2. sodium and oxygen gas react to yield sodium oxide

3. calcium carbonate and hydrogen sulfate react to yield calcium sulfate, water and carbon dioxide (CO_2).

4. hydrogen phosphate and sodium hydroxide react to produce sodium phosphate and water (hint: keep the phosphate as (PO_4) - do not separate!)

5. silver nitrate and nickel react to produce nickel nitrate and silver

6. aluminum sulfate and calcium hydroxide react to yield aluminum hydroxide and calcium sulfate

7. hydrogen iodide decomposes to produce hydrogen gas and iodine (which is I_2).

8. hydrogen peroxide decomposes to produce water and oxygen gas

9. zinc mixed with lead (II) nitrate yields zinc nitrate and lead

NAME_____ DATE_____
FRIENDLY CHEMISTRY

Lesson 21 Practice Page –2
Balancing Chemical Equations

Write and balance each chemical equation below. Ask if you need help!

1. potassium fluoride and chlorine gas react to yield potassium chloride and fluorine gas

2. strontium reacts with oxygen gas to yield strontium oxide

3. barium chloride and potassium bromide react to yield potassium chloride, barium and bromine gas

4. water and zinc chloride are the products when zinc oxide and hydrochloric acid (HCl) are mixed.

S211

5. carbon and aluminum oxide react to yield aluminum and carbon dioxide

6. barium oxide and carbon dioxide react to yield barium carbonate

7. magnesium carbonate decomposes to produce magnesium oxide and carbon dioxide

8. ammonium nitrate decomposes to yield nitrogen oxide (N_2O) and water

9. water decomposes to produce hydrogen gas and oxygen gas.

Lesson 22: Introduction to Stoichiometry

As you read the title of this chapter, you may be wondering, "What in the world are we going to discuss now?" Although the term, **stoichiometry** (pronounced stoy - key - om - uh - tree) may be a complex word to say or write, the meaning is quite simple. The root word stoichio- means the study of the elements, and the suffix -metry means to measure (i.e. meter). Stoichiometry is the study of measuring the elements. In our modern usage of the word, stoichiometry is the procedure of predicting the relative proportions of potential products to amounts of known reactant or vice versa.

In more simple terms, stoichiometry is a process to help you predict how much product you might expect to produce from a given amount of reactant. A good way to think about stoichiometry is to liken it to using recipes to cook food. For example, suppose you have a recipe for cookies like this:

<u>Good Cookies Recipe</u>

3 eggs

2 cups flour

Friendly Chemistry

1 cup butter
2 cups sugar

Mix all ingredients, bake for 10 minutes, yield 4 dozen cookies.

Assuming that this recipe is valid (actually works), you can use it to help you predict how many cookies you can prepare from a known amount of ingredient. For example, suppose that you have six eggs on hand (the recipe calls for three, right?). Provided you have plenty of all the other ingredients, you can predict that you can make eight dozen cookies (you have doubled the number of eggs, therefore provided you have enough of all the other ingredients, you can double the outcome of the recipe, ie 2 times 4 dozen = 8 dozen cookies).

Suppose you had 4 cups of butter available to make the same recipe and more than enough of all the other ingredients. What is the total number of cookies you could potentially make from these 4 cups of butter?

To answer this question, the first thing you do is to see how much butter is required for one batch of cookies. By looking at the recipe you see that it takes 1 cup butter to yield 4 dozen cookies. Four cups of butter is four times the needed amount, therefore you would be quadrupling the recipe, therefore you would expect to produce 16 dozen cookies.

Stoichiometry is very much like preparing a recipe. However, instead of using butter, eggs and flour, we use chemical compounds as our ingredients. Also, instead of using recipes, we use balanced chemical equations. And finally, instead of using conventional measurements like cups and teaspoons, chemists use moles and grams. Let's look at an example.

Example 1: Suppose you had 10 moles of ammonium nitrite. If you allowed this compound to decompose, water and nitrogen gas would result. How many moles of water might your expect if all 10 moles decomposes?

The first thing we need to do is write down the recipe, that is, the chemical reaction.

$$(NH_4)^{+1}{}_1(NO_2)_1 \longrightarrow H_2O_1 + N_2$$

Friendly Chemistry

It's very important to make sure that all compounds are written correctly. An error in this step can potentially cause major "headaches" in later steps.

After writing the chemical reaction, the next step is to balance the reaction. Use the process you learned in Lesson 21 to get this completed.

$$(NH_4)^{+1}{}_1(NO_2)_1 \longrightarrow H_2O_1 + N_2$$

Reactants	Products
N: 2	N: 2
H: 4	H: 2
O: 2	O: 1

Based upon this analysis, it appears that by placing a coefficient of 2 in front of the water on the product side of the reaction, the equation can be balanced.

$$(NH_4)^{+1}{}_1(NO_2)_1 \longrightarrow 2H_2O_1 + N_2$$

Reactants	Products
N: 2	N: 2
H: 4	H: 4
O: 2	O: 2

At this point its important to realize while there is no coefficient per se in front of any of the other members of this reaction, it is understood that there is 1 mole of each present. Writing these "1's" will help in the next step. We could rewrite the reaction like this:

$$1(NH_4)^{+1}{}_1(NO_2)_1 \longrightarrow 2H_2O_1 + 1N_2$$

Now we need to look back at our problem to find the two components of the reaction that we are being asked about. In this problem we are being asked to find the number of moles of water that we can produce from 10 moles of the ammonium nitrate. Find the ammonium nitrate and place a box around it. Do the same with the water. By doing so, it makes it a lot easier to keep things organized.

$$\boxed{1(NH_4)^{+1}{}_1(NO_2)_1} \longrightarrow \boxed{2H_2O_1} + 1N_2$$

The next step involves determining the **mole ratio** between these two reaction components. In this reaction for each 1 mole of ammonium nitrate, we produce 2 moles of water. Therefore the mole ratio between the ammonium nitrate and the water is a 1:2 ratio.

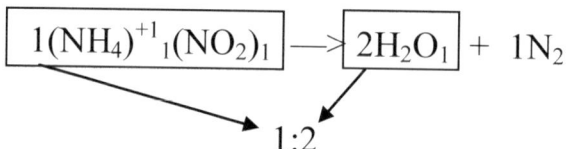

$$\boxed{1(NH_4)^{+1}{}_1(NO_2)_1} \longrightarrow \boxed{2H_2O_1} + 1N_2$$

$$1:2$$

This means that for every one mole of ammonium nitrate that we put into the reaction, we will produce two moles of water. If we were to put 2 moles in to the reaction, we'd get 4 moles of water produced from the reaction (which is like doubling the recipe, if we were making cookies!). If we were to put in 3 moles of ammonium nitrate, we'd get out 6 moles of water.

So, if we look back at the problem, we see that we are being asked to predict how many moles of water we could produce from 10 moles of ammonium nitrate. Based upon the ratio we just figured out, from 10 moles of ammonium nitrate we could produce 20 moles of water (we are multiplying the recipe 10 times).

Putting all of the steps in order will look like this:

1. Write the reaction. $(NH_4)^{+1}{}_1(NO_2)_1 \longrightarrow H_2O_1 + N_2$

2. Balance the reaction. $(NH_4)^{+1}{}_1(NO_2)_1 \longrightarrow 2H_2O_1 + N_2$

3. Add 1's where needed. $1(NH_4)^{+1}{}_1(NO_2)_1 \longrightarrow 2H_2O_1 + 1N_2$

4. Identify components asked about from the problem and box them.

$$\boxed{1(NH_4)^{+1}{}_1(NO_2)_1} \longrightarrow \boxed{2H_2O_1} + 1N_2$$

5. Determine the mole ratio between those two components in question.

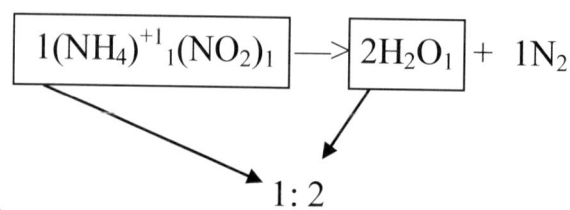

$$1:2$$

6. Apply the ratio to the quantity in the question.

$$10 \text{ moles of } (NH_4)^{+1}{}_1(NO_2)_1 \longrightarrow 20 \text{ moles of } H_2O_1$$

Let's look at another example where we are asked to predict the amount of product based upon a given amount of reactant.

Example 2: Potassium chloride will react with fluorine gas to produce potassium fluoride and chlorine gas. If you begin with 20 moles of fluorine gas, how many moles of potassium fluoride might you expect to be produced by the reaction?

Step 1: Write the chemical reaction. (your turn, go ahead and write it here).

Step 2: Balance the chemical reaction.

Step 3: Add 1's where necessary

Step 4: Identify the parts of the reaction you are being asked about.

At this point you should have something like this:

$$2 \text{ KCl} + \boxed{1 F_2} \longrightarrow \boxed{2 \text{ KF}} + 1 Cl_2$$

Step 5: Determine the mole ratio between the two components being asked about in the problem. In this case it will be:
$$1:2$$

Step 6: Apply the mole ratio to the given amount of reactant in the problem.

$$20 \text{ moles of } F_2 \longrightarrow 40 \text{ moles of KF}$$

Therefore, we can predict that from 20 moles of fluorine gas, we could produce 40 moles of potassium fluoride.

Before we try some more problems in the lesson practice pages, let's review what we've learned:

- Stoichiometry is the study of measuring the elements or in simpler terms, predicting the amount of product we can make with a known amount of reactant.

- An accurate balanced chemical equation (recipe) is needed first in solving a stoichiometry problem.

- The components in the reaction that we're concerned with need to be clearly identified. We put them into boxes.

- A mole ratio between those two components helps us predict the potential product from our given ingredient (reactant).

Time now for some practice. Turn to the Lesson 22 Practice pages and begin!

Name_____ Date_____

Friendly Chemistry

Lesson 22: Stoichiometry Practice
Predicting Moles Produced from a Given Amount of Ingredient

1. When hydrogen sulfate and sodium cyanide are mixed, hydrogen cyanide and sodium sulfate are produced. If you begin with 5 moles of sodium cyanide, how many moles of sodium sulfate might you produce from the reaction?

2. Zinc oxide and hydrochloric acid react to produce zinc chloride and water. If you begin with 24 moles of zinc oxide, how many moles of zinc chloride will you produce from this reaction?

3. Carbon and aluminum oxide react to produce aluminum and carbon dioxide (CO_2). If you have 13 moles of carbon, how many moles of aluminum might you expect to generate from this reaction? Carbon in this reaction will just be C with no subscript or charge.

4. When hydrogen gas and iodine gas are mixed, hydrogen iodide is produced. If you began with 0.5 moles of hydrogen gas, how many moles of hydrogen iodide will you potentially produce from this reaction?

5. Barium carbonate is the product of the reaction between barium oxide and carbon dioxide. If you begin with 28 moles of barium oxide, how many moles of barium carbonate could you produce from this reaction?

6. Magnesium and hydrogen chloride react to produce magnesium chloride and hydrogen gas. Suppose your boss requested you to prepare 65 moles of magnesium chloride from this reaction. When you looked on your shelf you saw that you had 120 moles of hydrogen chloride. Would this be enough to meet your boss' request? Prove your answer with calculations.

Name_____ Date_____

Friendly Chemistry

Lesson 22: Stoichiometry Practice –2
Predicting Moles Produced from a Given Amount of Ingredient

1. Magnesium chlorate will decompose to produce oxygen gas and magnesium chloride. If you began with 75 moles of magnesium chlorate, how many moles of magnesium chloride can you make from this reaction?

2. Iron (III) nitrate and lithium hydroxide react in a double replacement reaction to produce iron (III) hydroxide and lithium nitrate. If Gina began with 34 moles of lithium hydroxide, how many moles of lithium nitrate will she produce from this reaction?

3. James knew that zinc and sulfur would react to produce zinc sulfide. James had 38 moles of zinc. How many moles of the product could he make in his reaction?

4. Shelly took 5 moles of calcium and placed it into a beaker of bromine gas. Calcium bromide resulted from the reaction that consequently took place. How much calcium bromide did she make from her 5 moles of calcium?

5. Water will decompose into hydrogen gas and oxygen gas under appropriate conditions. If you begin with 7.8 moles of water, how many moles of hydrogen gas could you produce from this reaction?

6. Lead (II) nitrate and zinc readily react to produce zinc nitrate and lead. How many moles lead could you produce from 40 moles of zinc?

7. Lye (sodium hydroxide) and hydrogen sulfate react to produce sodium sulfate and water. If you begin with 4.5 moles of lye, how many moles of water could you produce from this reaction?

Lesson 23: Predicting Grams of Product Produced in a Reaction

In the last lesson, you learned how to set up a stoichiometry problem to help you predict how many moles of a product you could make from a given amount of ingredient. You learned how to first write the reaction, balance it, box the components in the reaction in which you were interested, find the mole ratio and finally apply the ratio to the given amount of ingredient/reactant in the reaction to predict the number of **moles** produced by the reaction.

In this lesson, we will go one step further. Instead of stopping with the amount of moles produced by a reaction, we will calculate how many **grams** we can produce from the reaction. Of course, first we will have to determine how many moles of product are made by the reaction. Then we will convert that answer in moles to **grams**.

Let's look at an example from the last lesson (#22).

Example 1: Suppose you had 10 moles of ammonium nitrite. If you allowed this compound to decompose, water and nitrogen gas would result. How many **grams** of water might your expect if all 10 moles decomposes?

S223

This was our solution to this example:

$$(NH_4)^{+1}{}_1(NO_2)_1 \longrightarrow 2H_2O_1 + N_2$$

$$1(NH_4)^{+1}{}_1(NO_2)_1 \longrightarrow 2H_2O_1 + 1N_2$$

$$1:2$$

$$10 \text{ moles of } (NH_4)^{+1}{}_1(NO_2)_1 \longrightarrow 20 \text{ moles of } H_2O_1$$

We found that 20 moles of H_2O could be produced in this reaction. In this lesson we will take the process one step further. Knowing you could make 20 moles of water is indeed useful, but knowing how many grams of water produced would be even more useful!

Let's look at how you might go about this. Recall from Lesson 17 where you learned how to convert moles of a compound into **grams** of a compound (finding the formula weight). You learned that the atomic mass values found on the periodic table of elements told you how many grams one mole of a particular element weighed. We will use this same process to convert moles of a product to grams of a product.

In this example, the reaction produced 20 moles of water. In order to convert these moles of water into grams, first we must find the formula weight of water (the number of grams in one mole of water). Once we know the formula weight, we simply multiply that by the number of moles produced. In other words, if one mole of water weighs "this much," then 20 moles must weigh 20 times "that much."

So, let's find the formula weight of water.

$$H_2O$$

H: 2 moles x 1 gram/mole =   2 grams
O: 1 mole x 16 grams/mole = 16 grams
                                                    18 grams

From these calculations, we can say that one mole of water weighs 18 grams. So, now we multiply the number of moles produced in the reaction time 18 grams.

$$20 \text{ moles } H_2O \times 18 \text{ grams/mole} = 360 \text{ grams of } H_2O$$

Friendly Chemistry

So, from our reaction we can say that 20 moles of water were produced which is equivalent to 360 grams of water.

Let's look at another example: Barium oxide and carbon dioxide react to produce barium carbonate. If you begin with 5 moles of barium oxide, how many **grams** of barium carbonate can you produce from this reaction?

Write the balanced equation here:

The moles ratio is:

From the 5 moles of barium oxide, _____ moles of barium carbonate are produced.

Now, we need to determine the formula weight of barium carbonate. Do that here:

Ba: x   =

C: x   =

O: x   = _____
Formula weight   = _____

So, in our reaction we produced _____ moles of barium carbonate. We take moles we produced x the formula weight of barium carbonate:

_____ x _____ = _____

<u>Check your work by looking at the answers here:</u>
Balanced equation: 1 Ba(O) + 1 CO$_2$ —> 1 Ba(CO$_3$)
Mole ratio: 1:1
5 moles BaO —> 5 moles Ba(CO$_3$)
Formula weight of Ba (CO$_3$) = 197 grams

5 moles Ba(CO$_3$) x 197 grams/mole = **985 grams Ba(CO$_3$)**

S225

To review this process, first we:
- Wrote and balanced the chemical equation.
- Then we determined the mole ratio.
- Next we applied the mole ratio to the reactant in question to find the number of moles produced.
- And, finally, we found the number of grams per mole of the product (formula weight) and multiplied it by the number of moles produced.

Practice this procedure by completing the practice problems on the following pages..

Name_____ Date_____
Friendly Chemistry

Lesson 23: Stoichiometry Practice –1
Predicting Grams Produced from a Given Amount of Ingredient

1. When hydrogen sulfate and sodium cyanide are mixed, hydrogen cyanide and sodium sulfate are the produced. If you begin with 25 moles of sodium cyanide, how many **grams** of sodium sulfate might you produce from the reaction?

2. Zinc oxide and hydrochloric acid react to produce zinc chloride and water. If you begin with 14 moles of zinc oxide, how many **grams** of zinc chloride will you produce from this reaction?

3. Carbon and aluminum oxide react to produce aluminum and carbon dioxide (CO_2). If you have 1.4 moles of carbon, how many **grams** of aluminum might you expect to generate from this reaction?

4. When hydrogen gas and iodine gas are mixed, hydrogen iodide is produced. If you began with 74 moles of hydrogen gas, how many **grams** of hydrogen iodide will you potentially produce from this reaction?

5. Barium carbonate is the product of the reaction between barium oxide and carbon dioxide. If you begin 0.4 moles of barium oxide, how many **grams** of barium carbonate could you produce from this reaction?

6. Magnesium and hydrogen chloride react to produce magnesium chloride and hydrogen gas. If you begin with 4.6 moles of magnesium, how many **grams** of magnesium chloride can you produce from this reaction?

Name_____ Date_____
Friendly Chemistry

Lesson 23: Stoichiometry Practice –2
Predicting Grams Produced from a Given Amount of Ingredient

1. Magnesium chlorate will decompose to produce oxygen gas and magnesium chloride. If you began with 7.05 moles of magnesium chlorate, how many **grams** of magnesium chloride can you make from this reaction?

2. Iron (III) nitrate and lithium hydroxide react in a double replacement reaction to produce iron (III) hydroxide and lithium nitrate. If Morris began with 13 moles of lithium hydroxide, how many **grams** of lithium nitrate will she produce from this reaction?

3. Jon knew that zinc and sulfur would react to produce zinc sulfide. Jon had 38 moles of zinc. How many **grams** of the product could he make in his reaction?

4. Tori took 0.97 moles of calcium and placed it into a beaker of bromine gas. Calcium bromide resulted from the reaction that consequently took place. How many **grams** of calcium bromide did she make from her 0.97 moles of calcium?

5. Water will decompose into hydrogen gas and oxygen gas when an electric current is applied to it. If you begin with 24 moles of water, how many **moles** of hydrogen gas could you produce from this reaction?

6. Lead (II) nitrate and zinc readily react to produce zinc nitrate and lead. How many **grams** of lead could you produce from 9 moles of zinc?

7. Lye (sodium hydroxide) and hydrogen sulfate react to produce sodium sulfate and water. If you begin with 4.5 moles of lye, how many **grams** of water could you produce from this reaction?

8. When lithium metal is exposed to oxygen gas, a rapid reaction takes place resulting with the production of lithium oxide. If you begin with 3.2 moles of lithium, how many **grams** of lithium oxide can you expect to produce from the reaction?

Lesson 24: Predicting Grams of Product from Grams of Reactant

In the last lesson, you learned how to set up a stoichiometry problem to help you predict how many grams of a product you could make from a given amount of moles of reactant. In this lesson you will learn how to predict the amount of **grams** of product created in a reaction from a **given number of grams of reactant**.

Lets look at an example:

Example 1: Suppose you had 350 **grams** of ammonium nitrite. If you allowed this compound to decompose, water and nitrogen gas would result. How many **grams** of water might your expect if all 350 grams decomposes?

Begin your problem by writing the balanced chemical equation.

$$(NH_4)^{+1}{}_1(NO_2)_1 \longrightarrow 2H_2O_1 + N_2$$

Friendly Chemistry

Locate the compounds in question.

$$\boxed{1(NH_4)^{+1}{}_1(NO_2)_1} \longrightarrow \boxed{2H_2O_1} + 1N_2$$

1:2

At this point in these types of stoichiometry problems we apply the mole ratio to the reactant in question to allow us to predict the number of moles (or subsequently grams) produced by the reaction. However, in this problem we cannot do this. We have been given an amount of reactant in **grams** (and not in **moles**). In order to be able to apply the amount of ingredients to the mole ratio, we must first convert the number of **grams** we've been given to number of **moles**.

In the problem, we've been given 350 grams of ammonium nitrite. In order to convert these grams into moles we must first find the number of grams in one mole of ammonium nitrite. Hopefully, you remember that finding the formula weight tells you this answer.

Formula weight of ammonium nitrite:

N: 2 x 14 grams/mole = 28 grams

H: 4 x 1 gram/mole = 4 grams

$\underline{O: 2 \times 16 \text{ grams/ mole} = 32 \text{ grams}}$
Formula weight = 64 grams

This calculation tells us that in one mole of ammonium nitrite weighs 64 grams. Therefore, to find how many moles this is equivalent to, we can set up the following mathematical relationship:

$$\frac{1 \text{ mole } NH_4(NO_3)}{64 \text{ grams}} = \frac{X \text{ moles}}{350 \text{ grams}}$$

64X = 350
X = 5.46 moles of $NH_4(NO_2)$

Now, that we've converted the grams of reactant we've been given to moles, we can apply it to the mole ratio in the problem.

The mole ratio is:
1:2

Therefore:
5.46 moles of reactant —> 10.92 moles of product
Or
5.46 m $NH_4(NO_2)$ —> 10.92 m H_2O

At this point our answer is 10.92 moles of water. If we look back at our problem, we see that we've been asked to predict the number of **grams of water** produced by the reaction. Consequently, we now need to convert our answer in moles to grams (which is exactly what you learned in the last lesson.)

1 mole of H_2O = 18 grams
(this was just finding the formula weight of water)

From our problem, we found we produced 10.92 moles of water. Therefore:

10.92 m H_2O x 18 grams/mole = 196.56 grams H_2O

Hopefully, you've realized that converting our grams of reactant to moles of reactant is a process we've learned earlier. We are now applying what you learned previously to stoichiometry.

Let's review the steps we learned in this lesson:
- Write and balance the chemical reaction.
- Locate compounds in question.
- Determine the mole ratio.
- Convert given **grams** of reactant to **moles** of reactant.
- Apply moles of reactant to mole ratio to get moles of product.
- Convert moles of product to grams of product.
- Be happy because you've completed the problem!

Continue with Lesson 24 Practice Problems.

Name_____ Date_____
Friendly Chemistry

Lesson 24: Stoichiometry Practice –1
Predicting Grams Produced from a Given Amount of Grams of Reactant

1. Carbon and aluminum oxide react to produce aluminum and carbon dioxide (CO_2). Suppose that Tony has 48 **grams** of carbon, how many **grams** of aluminum might he expect to generate from this reaction?

2. Zinc oxide and hydrochloric acid react to produce zinc chloride and water. If you begin with 78 grams of zinc oxide, how many grams of zinc chloride will you produce from this reaction?

3. When hydrogen sulfate and sodium cyanide are mixed, hydrogen cyanide and sodium sulfate are the produced. Suppose Chuck has 100 grams of hydrogen sulfate. How many grams of hydrogen cyanide might he expect to produce from this reaction?

4. Mary had 456 grams of barium oxide. She knew that barium carbonate is the product of the reaction between barium oxide and carbon dioxide. If she used all of her barium oxide in the reaction, how many grams of barium carbonate might she expect to produce?

5. When hydrogen gas and iodine gas are mixed, hydrogen iodide is produced. Sara's boss needed her to prepare some hydrogen iodide for a procedure they were doing later that day. If Sara had 3000 grams of iodine gas on hand, how many grams of hydrogen iodide could she make for her boss?

6. Magnesium and hydrogen chloride react to produce magnesium chloride and hydrogen gas. If you begin with 4.6 **moles** of magnesium, how many grams of magnesium chloride can you produce from this reaction?

Friendly Chemistry

Name_____ Date_____
Friendly Chemistry

Lesson 24: Stoichiometry Practice –2
Predicting Grams Produced from a Given Amount of Grams of Reactant

1. Franklin found in his chemistry manual that magnesium chlorate will decompose to produce oxygen gas and magnesium chloride. If he began with 3 kg of magnesium chlorate, how many kilograms of magnesium chloride could he make from this reaction? (one kilogram = 1000g and atomic mass values are always given in grams).

2. Iron (III) nitrate and lithium hydroxide react in a double replacement reaction to produce iron (III) hydroxide and lithium nitrate. If Sheila began with 74 grams of lithium hydroxide, how many grams of lithium nitrate will she produce from this reaction?

3. Jeremy knew that zinc and sulfur would react to produce zinc sulfide. If Jeremy had half a bottle of zinc pellets available for his reaction, how many grams of zinc sulfide could he produce? The bottle originally held 1500 grams of zinc.

4. Billy took 350 grams of calcium and placed it into a beaker of bromine gas. Calcium bromide resulted from the reaction that consequently took place. His instructor needed at least 400 grams of calcium bromide for a class project. Could Billy make enough with this 350 grams? Show proof of your answer with calculations.

5. Hydrogen gas and oxygen gas are the products of the decomposition of water. If you took 50 mls of water and allowed it all decompose into hydrogen and oxygen gas, how many grams of oxygen gas might you yield from this reaction? (1 ml of water = 1 gram of water).

6. Lead (II) nitrate and zinc readily react to produce zinc nitrate and lead. How many grams of lead could you produce from 5 grams of lead (II) nitrate?

7. When lithium metal is exposed to oxygen gas, a rapid reaction takes place resulting with the production of lithium oxide. If you begin with 6700 grams of lithium, how many grams of lithium oxide can you expect to produce from the reaction?

Lesson 25: Do I Have Enough Reactant?

Up to this point with our stoichiometry problems, we've been predicting the amount of **product** we could potentially produce from a **given amount of reactant**. In this lesson, we are going to switch gears a bit. Let's go way back to our analogy of making cookies. Suppose you've been asked to prepare six dozen chocolate chip cookies. You have a recipe that makes 2 dozen cookies so you know you'll be needing to triple the recipe. Your next logical step would be to determine if you have enough ingredients on hand to allow you to make this triple batch. You know that if you triple the amount of desired product (cookies in this case) you will need to triple the amount of each ingredient that goes into the recipe.

In this lesson you will learn how to work problems using chemical compounds in just the same way. You will be asked (theoretically) to prepare certain amounts of chemical compounds from given chemical reactions (recipes). Based upon these desired amounts of products, you will be asked to determine how much reactant (ingredient) it will take to prepare those amounts.

Lets look at an example:

Example 1: Suppose you are being asked to prepare 5 moles of nitrogen gas from the following reaction: ammonium nitrite will decompose to produce water and nitrogen gas. How many moles of ammonium nitrite will you need to prepare these 5 moles of nitrogen gas?

Begin solving this problem by writing and balancing the chemical equation.

$$(NH_4)^{+1}{}_1(NO_2)_1 \longrightarrow 2H_2O_1 + N_2$$

Locate the compounds in question.

$$\boxed{1(NH_4)^{+1}{}_1(NO_2)_1} \longrightarrow 2H_2O_1 + \boxed{1N_2}$$

Like we've done in the past, we need to determine the mole ratio between our desired product and reactant we are being asked to find. In this case, we have a 1:1 ratio or in other words, in order to make 1 mole of nitrogen gas, it takes 1 mole of ammonium nitrite.

In our problem we were asked to find out how many moles of ammonium nitrite it would take to make 5 moles of nitrogen gas. It stands to reason based upon the mole ratio determined above, that it would take 5 moles of ammonium nitrite to make 5 moles of nitrogen gas (1:1 ratio).

You can see that this process is very similar to the steps you used in predicting amounts of products from a given amount of reactant. In this case, we compare the amount of product we would like to make to the amount of needed ingredient.

Let's look at another example:

Example 2: Potassium perchlorate decomposes to produce potassium chloride and oxygen. How many moles of potassium perchlorate would it take to produce 20 moles of oxygen gas?

Begin by writing and balancing the chemical reaction:

$$1K(ClO_4) \longrightarrow 1KCl + 2O_2$$

Identify the parts of the reaction that you are concerned with.

$$\boxed{1K(ClO_4)} \longrightarrow 1KCl + \boxed{2O_2}$$

Determine the mole ratio between the two components. In this reaction we see that for every one mole of potassium perchlorate we put into the reaction, we produce 2 moles of oxygen gas or a 1:2 ratio. Another way to think about this ratio is that for every two moles of oxygen gas we produce, one mole of potassium perchlorate is necessary. Or yet another way to think about this relationship is that it takes half as many moles of potassium perchlorate to make a desired number of moles of oxygen gas

If you look back now at the problem you will see that we were asked to figure out how many moles of potassium perchlorate it would take to make 20 moles of oxygen gas. Based upon the mole ratio, we can now say it would take 10 moles of potassium perchlorate to yield 20 moles of oxygen gas.

In the last two lessons (23 and 24) you learned that learning to predict moles of product was useful, but being able to convert those results into **grams** was even more meaningful! We can apply those same techniques when we find out the amount of needed ingredient necessary to produce a desired amount of product.

If we look back at our example above (Example 2), we found that it took 10 moles of potassium perchlorate to yield our desired amount of 20 moles of oxygen gas. Knowing we would need 10 moles of potassium perchlorate is good, but knowing how many grams this is equivalent to is even more useful. In order to make the conversion from moles to grams, recall that we multiply the number of moles we need by the number of grams present in one mole of that compound (also known as the formula weight!).

In this case, we'll need to find the formula weight of potassium perchlorate and then multiply that result times the number of moles we need. So we need to find the formula weight of potassium perchlorate first:

$$K: 1 \times 39 = 39 \text{ grams}$$
$$Cl: 1 \times 23 = 23 \text{ grams}$$
$$\underline{O: 4 \times 16 = 64 \text{ grams}}$$
$$\text{Formula weight} = 126 \text{ grams}$$

Therefore, each mole of potassium perchlorate weighs 126 grams. Ten moles of the compound would then have the mass of 1260 grams (10 moles X 126 grams/mole).

So to make the desired amount of oxygen gas (20 moles), we'd need to put 1260 grams of potassium perchlorate into the reaction.

To review this process of finding the amount of ingredient necessary to produce a desired amount of product we:
- First, wrote and balanced the chemical reaction.
- Next, we identified the product we desired and the reactant in question.
- We then determined the mole ratio between the two compounds in question.
- Next, we applied the mole ratio to the two compounds in question which gave us the resulting amount of needed reactant in moles.
- Finally, we converted the needed reactant given in moles to grams making it a more practical result. We did this by multiplying the number of needed moles by the number of grams found in one mole of the reactant (formula weight.)

Continue by working the problems on the Lesson 25 practice pages.

Name_____ Date_____
Friendly Chemistry

Lesson 25: Stoichiometry Practice –1
How Much Reactant Do I Need?

1. When zinc is dropped into sulfuric acid (hydrogen sulfate), zinc sulfate and hydrogen gas are produced. If you needed to produce 40 moles of zinc sulfate, how many moles of zinc would this require?

2. Iron and copper (II) sulfate react to produce iron (II) sulfate and copper. If you needed to produce 34 moles of copper from this reaction, how many moles of iron would be required?

3. Ammonium chloride and calcium hydroxide react to yield calcium chloride, ammonia (NH_3), and water. If your boss asked you to prepare 45 moles of calcium chloride from this reaction, how many **grams** of ammonium chloride would you need to have ready?

4. Lead (II) sulfide can be produced from the reaction of hydrogen sulfide and lead (II) chloride. Hydrogen chloride is a byproduct of the reaction. If you needed to prepare 400 moles of lead (II) sulfide, how many grams of EACH reactant will you need? Hint: determine two separate mole ratios.

5. Ammonia (NH_3) reacts with hydrogen sulfide to produce ammonium sulfide. If you need 2.3 moles of ammonium sulfide, how many grams of ammonia would you need?

Name_____ Date_____

Friendly Chemistry

Lesson 25: Stoichiometry Practice –2
How Much Reactant Do I Need?

1. Potassium chloride, when exposed to oxygen gas, produces potassium perchlorate. If you need to produce 13 moles of potassium perchlorate, how many grams of potassium chloride would you need?

2. When iron is exposed to oxygen gas, rust (Iron (III) oxide) forms. If you begin with 25 grams of iron, how many grams of iron (III) oxide can you **produce** from this reaction?

3. Ammonium sulfide decomposes to produce ammonia and hydrogen sulfide. If you begin with 3 pounds of ammonium sulfide, how many grams of hydrogen sulfide might you expect to **produce** from this reaction? Note that 1 pound = 454 grams!

4. If bromine gas is bubbled through potassium iodide, potassium bromide and iodine crystals are produced. Suppose you need to make 100 grams of iodine crystals. How many grams of potassium iodide do you need?

5. When aluminum is combined with sulfuric acid (hydrogen sulfate), aluminum sulfate and hydrogen gas are produced. If you need to produce 30 moles of aluminum sulfate, will 400 grams of aluminum be enough? Show proof of your answer with calculations.

Lesson 26: Mixing Compounds with Water to Make Solutions

Up to this point in our discussion of chemical reactions, we have been assuming all of our reactants were in the dry form such as powders or crystals. However, some compounds are actually mixed with water or other solvents to make them more willing to react. Plus, some compounds are safer and easier to handle if they are mixed with water. However, it's still important to know how much of the compound has been mixed into the solution before we use the solution in reactions. Like before, we will still use moles as the unit of measure in our reactions, but because these moles in some reactants will now be mixed into water, we use the measurement called **molarity (M).**

Before we continue, lets take a moment to learn some names of the parts of a solution. The compound that we are mixing into the water is called the **solute**. For example, if we are mixing sugar into our ice tea, the sugar will be called the solute. The water part of the solution is known as the **solvent**. In our ice tea example, the tea would be considered the solvent. So, solute is what is being added or mixed into the solvent (the stuff doing the dissolving). Now, lets look back at the term molarity.

Molarity tells how many moles of the solute has been mixed into one liter of the solvent (usually water in most cases). For example, a solution that is labeled a 1 M or 1 Molar solution means that one mole of the compound has been mixed into 1 liter of water. A 2 M or 2 Molar solution means that 2 moles of the compound has been mixed into one liter of water. A 3 M solution means that 3 moles of the compound has been mixed into 1 liter of water. Obviously, the greater the molarity of the solution, the more concentrated the solution would be. A 12 M solution would be twice as concentrated as a 6 M solution.

Consider how these solutions are made. Suppose you needed to prepare a 1 M solution of table salt (sodium chloride). A 1 molar solution means one mole of sodium chloride has been mixed into 1 liter of water. In order to make this solution, you will need to get 1 mole of the sodium chloride. Since we don't have scales that measure in moles, we need to calculate the number of grams in one mole of sodium chloride. Hopefully, you'll readily recall that this is a formula weight for the sodium chloride. By doing your calculations, you should find that one mole of sodium chloride weighs approximately 58 grams. A 1 molar solution would therefore have 58 grams of the sodium chloride mixed into 1 liter of water.

Consider how many grams of sodium chloride would be required to make a two molar (2 M) solution. If a 1 M solution requires 58 grams, it should be obvious that a 2 M solution would require 2 times as much sodium chloride or 2 x 58 grams = 116 grams of NaCl.

Try this problem: Suppose you were asked to prepare a 5 M solution of potassium hydroxide. How many grams of potassium hydroxide powder would you need?

Like in the example above, we first need to find the number of grams in 1 mole of potassium hydroxide.

$$K_1(OH)_1$$
K: 1 x 39g = 39 g
O: 1 x 16 = 16 g
H: 1 x 1 = 1 g
56 g in one mole

Therefore, a 5 M solution would be prepared by mixing 5 times the grams found in one mole or 5 x 56 grams = 280 grams of K(OH).

Friendly Chemistry

Try another problem: Suppose you were asked to prepare 2 L of a 3 molar calcium chloride solution. Where do you start? _____

Good, find the formula weight for calcium chloride. Do your work here:

So, the formula weight for calcium chloride is _____ grams. (If you didn't get 110 grams, check back over your work!) One mole of calcium chloride weighs 110 grams.

Now, the problem asks you to make a 3 M solution. Therefore, 3 x 110 grams = 330 grams. But, remember this is how much calcium chloride is needed to be added to **one** liter of water. The problem asks you to prepare **2 liters** of the solution. Since one liter will require 330 grams, it should be intuitive that to prepare 2 liters of solution, it will take 2 x 330 grams or 660 grams of calcium chloride. Therefore, in order to prepare 2 liters of a 3 M calcium chloride solution, we would need to use 660 grams of calcium chloride.

Let's look at one final example. Suppose you worked in a college chemistry lab and your boss asked you to prepare 5 liters of a 12 molar barium oxide solution. How many grams of barium oxide would you need to get from your stock supply?

Where do you begin? Hopefully, you recall to find the formula weight of barium oxide first. Do your work here:

Based upon these calculations, a 1 M solution would require _____ grams.

Therefore, a 12 M solution of barium oxide would require _____ grams.

Since you need five liters of this solution, now you will take 5 times the amount needed for one liter: 5 x _____ grams = _____ grams.

Correct answers can be found on the next page.

To review, in this lesson we learned that some compounds are more readily handled if they are mixed into solutions. We learned that we still had to be aware of how much of the compound was mixed into the solution. As in the past lessons, we learned

S249

we will still use moles to measure amounts of chemicals and that the amount of solute mixed into the solvent was called molarity (M). A 1 M solution has one mole of solute mixed into 1 liter of solvent. In order to prepare a solution of a desired molarity (M), we found the number of grams per 1 mole of solvent (formula weight).. Finally, if we needed more than one liter of the solution, we learned to multiply the amount needed in one liter, times the number of liters desired.

Continue to practice these new concepts by completing the Lesson 26 Practice Pages.

Answers for the barium oxide (BaO) example:

Formula weight of barium oxide = 153 grams.

A 12 M solution would require 1836 grams.

And 5 liters of a 12 M solution would require 9180 grams.

Friendly Chemistry

Name_____ Date_____
Friendly Chemistry

Lesson 26: Preparing Solutions –1

Read each problem below. Work slowly and carefully.

1. Tony was asked to prepare a 3 M potassium chloride solution. How many grams of potassium chloride does Tony need?

2. Francis needed to prepare a 5.5 M sodium sulfate solution. How many grams of sodium sulfate does she need?

3. Harry was told to make a 0.5 M solution of nickel chromate. How many grams does of nickel chromate does Harry need?

4. If you needed to prepare 2 liters of a 6 M calcium carbonate solution, how many grams of calcium carbonate do you need?

5. Shorty needed to prepare 10 liters of a 4 M lithium acetate solution. How many grams of lithium acetate does he need?

6. Lisa was asked to prepare 0.5 liters of a 2 M iron (III) chloride solution. How many grams of iron (III) chloride does she need?

7. Tommy used 513 grams of calcium phosphate to prepare one liter of solution. What was the molarity of this solution?

8. Sherry added 720 grams of copper (II) sulfite to one liter of water. What was the molarity of the solution she made?

Name_____ Date_____

Friendly Chemistry

Lesson 26: Preparing Solutions –2

1. Frank needed to make 4 liters of a 3 M potassium permanganate solution. How much potassium permanganate does Frank need?

2. Trish had 294 grams of copper (II) hydroxide. If she added this to 1 liter of solvent, what would the resulting molarity of the solution be?

3. Dan was asked to prepare 10 liters of a 0.1 M hydrogen chloride solution. How many grams of hydrogen chloride does he need?

4. Toni put 35 grams of barium iodide into 1 liter of water. What was the resulting molarity of the solution?

5. Billy put 360 grams of calcium sulfite into one liter of water. What was the molarity of the resulting solution?

6. Charlotte and Perry were each asked to prepare a 0.6 M solution of magnesium cyanide. Charlotte said they needed 45.6 grams. Perry said they needed 76 grams. Who was correct? Or were they both wrong? Show proof of your answer.

7. Antonio was asked to prepare a 17.5 M ammonium hypochlorite solution. How many grams of the compound should he get?

8. Sheila told Tom she was going to make a 0.1 M sodium chloride solution. She said she needed to get 5.8 grams of sodium chloride. Was she correct? Show proof with calculations.

Lesson 27: Incorporating Molarity into Stoichiometric Problems

In Lesson 26 you learned how to prepare solutions of specific molarities. You learned that a 1 M solution contained 1 mole of solute dissolved into 1 liter of solvent. You also learned how to prepare solutions of varying molarities. Now that you've learned this technique, the next step is learning how to use those solutions to carry out chemical reactions.

Earlier you learned the steps of stoichiometry to help you predict amounts of products created by reactions and amounts of reactants necessary to produce a desired amount of product. In each of those stoichiometric problems you solved, you either began with a certain amount of grams or moles of reactant or desired product. In this lesson you will learn how to take a volume of solution of known molarity and place it into a chemical reaction where you will predict the amount of potential product.

Let's look at an example:

Example 1: When sodium hydroxide is mixed with hydrogen chloride, sodium chloride and water result. Suppose you have been given 100 ml of a 5 M solution of sodium hydroxide. How many grams of sodium chloride could you yield from this reaction?

First, let's begin by writing our chemical reaction.

$$Na(OH) + HCl \longrightarrow NaCl + H_2O$$

Then, balance the reaction.

$$1Na(OH) + 1HCl \longrightarrow 1NaCl + 1H_2O$$

Locate the reaction components we are interested in and place them in boxes.

$$\boxed{1Na(OH)} + 1HCl \longrightarrow \boxed{1NaCl} + 1H_2O$$

Determine the mole ratio.

$$1 \text{ m Na(OH)} : 1 \text{ m NaCl}$$

This ratio says one mole of sodium hydroxide will yield one mole of sodium chloride. If we look back at our problem, we've been given 100 ml of a 5 M solution. In order to apply the mole ratio to the reactant in question, we must first figure out how many moles of sodium hydroxide we've been given.

Let's look at the solution: it is a 5 M solution meaning that 5 moles of sodium hydroxide have been mixed into 1 liter of solvent. By doing a formula weight of sodium hydroxide we find that one mole of sodium hydroxide is equivalent to 40 grams. A 5 M solution would therefore have 5 x 40 grams/mole = 200 grams dissolved in one liter of solvent. However, in this problem we aren't using one whole liter of solution.

If we assume the sodium hydroxide is equally distributed throughout the solution, we can then determine how many grams of the sodium hydroxide that we could find in 100 ml of the solution. An easy way to do this is like this: In 1000 ml (1 liter) of the solution we know there are 200 grams of sodium hydroxide. Then our question becomes, how many grams would there be in 100 mls.

Set up equivalent fractions like these. We know there are 200 grams of sodium hydroxide in 1000 mls of solution, so how many grams (x) would there be in 100 mls of solution?

$$\frac{200 \text{ grams NaOH}}{1000 \text{ ml}} = \frac{x \text{ grams}}{100 \text{ ml}}$$

Cross multiply and solve for x.

$$1000x = 20,000$$

$$x = 20 \text{ grams}$$

Therefore in our 100 mls of solution, we would actually be putting 20 grams of sodium hydroxide into the reaction.

To continue the problem, as you have done in past stoichiometry problems, you'll need to convert these grams we've calculated into moles before we can apply the mole ratio.

In order to do this, we again need to know the formula weight of sodium hydroxide (number of grams in one mole). From our earlier calculations, we know that one mole of sodium hydroxide is equivalent to 40 grams. So…..

$$\frac{1 \text{ m NaOH}}{40 \text{ g}} = \frac{x \text{ m}}{20 \text{ g}}$$

Cross multiply and solve for x.
$$40x = 20$$
$$x = 0.5 \text{ moles}$$

So, now we can say that we're actually putting 0.5 moles of sodium hydroxide into the reaction.

Recall from above that our mole ratio was 1:1. If we put 0.5 moles of sodium hydroxide in to the reaction, we'll produce 0.5 moles of sodium chloride. Finally, we just need to convert this result (0.5 m NaCl) into grams of NaCl. Once again, we'll need the number of grams found in one mole of NaCl. Calculating the formula weight, we find that one mole of NaCl equals 58 grams.

Therefore 0.5 moles NaCl x 58 grams/mole = 29 grams NaCl. So, from the 100 mls of 5 M sodium hydroxide, we can predict that we will produce **29 grams of sodium chloride from the given chemical reaction**.

Let's look at another example:

Example 2: Mike had 230 mls of 3 M hydrogen sulfate (sulfuric acid). If he added several aluminum cans to this solution, aluminum sulfate and hydrogen gas would be produced. How many grams of aluminum sulfate could he make from his 230 mls of 3 M hydrogen sulfate?

First, write the reaction:

$$H_2(SO_4)_1 + Al \longrightarrow Al_2(SO_4)_3 + H_2$$

Then balance and locate the compounds we are interested in.

$$\boxed{3\ H_2(SO_4)_1} + 2Al \longrightarrow \boxed{1Al_2(SO_4)_3} + 3\ H_2$$

Determine the mole ratio:

$$3 : 1$$

Now, go back to the problem. We've been given 230 mls of a 3 M solution of hydrogen sulfate. We need to figure out how many moles of hydrogen sulfate this is equivalent to in order to apply it to the mole ratio.

We know that in a 3 M solution, there are 3 moles of hydrogen sulfate. One mole of hydrogen sulfate equals how many grams? How can you determine that? If you said, "Find the formula weight!" that's correct!

So, after calculation we find the formula weight of hydrogen sulfate to be 98 grams. A 3 M solution would therefore have 3 x 98 grams/mole – 490 grams per liter.

Now we can set up two equivalent fractions:

$$\frac{294\ grams\ H_2(SO_4)}{1000\ ml} = \frac{x\ grams}{230\ ml}$$

These fractions say: "we know there are 490 grams of H_2SO_4 in 1000 ml, so how many grams (x) would there be in 230 mls. By cross-multiplying and solving for x, we get:

$$x = 67.62\ grams\ H_2(SO_4)$$

Friendly Chemistry

From these calculations we now know that we are starting this reaction with 67.62 grams of $H_2(SO_4)$. From this point the process of solving the problem is just like any of the other stochiometry problem you've done in the past.

We balanced the equation above. Here it is written again.

$$3H_2(SO_4)_1 + 2Al \longrightarrow 1Al_2(SO_4)_3 + 3H_2$$

And, we also found the ratio to be: 3:1

Because we are working with grams of ingredient, recall that before we can apply the mole ratio, we must convert these grams into moles of ingredient.

$$\frac{1 \text{ mole } H_2(SO_4)}{98 \text{ grams}} = \frac{x \text{ moles}}{67.62 \text{ grams}}$$

$$98x = 67.62$$
$$x = 0.69 \text{ moles of } H_2(SO_4)$$

So, our 112.7 grams of hydrogen sulfate is equal to 1.15 moles of hydrogen sulfate. Now, we can apply this value to the mole ratio.

$$3 : 1$$

$$0.69 \text{ m } H_2(SO_4) \longrightarrow 0.23 \text{ m } Al_2(SO_4)_3$$

At this point, we can say that from the 230 mls of 3 M hydrogen sulfate, **we can produce 0.23 m of aluminum sulfate**. However, we're not quite done yet! The problem asks us to determine the number of **grams** of aluminum sulfate that we can yield from the reaction. We need to convert these 0.23 moles we've produced into **grams** of aluminum sulfate that we've produced. In order to do this, we'll need a formula weight for aluminum sulfate:

Aluminum sulfate
$Al_2(SO_4)_3$

Al: 2 x 27g = 54 g

Friendly Chemistry

$$S: 3 \times 32 \text{ g} = 96 \text{ g}$$
$$\underline{O: 12 \times 16\text{g} = 192 \text{ g}}$$
$$\text{Formula wt} = 342 \text{ g}$$

So, 0.23 m x 342g per mole = 78.66 grams $Al_2(SO_4)_3$

So, finally, we can say that from 230 mls of 3 M hydrogen sulfate, we can predict that our yield will be 78.66 grams of aluminum sulfate.

So, to review:
- First, write and balance the reaction. Locate the reaction components you are concerned with. Determine the mole ratio.
- Find in the problem the amount (ml) and molarity (M) of ingredient that you are given.
- Convert mls of specified reactant into grams of reactant. Then convert those grams into moles.
- Apply the mole ratio to determine moles of product produced.
- Convert moles of product produced into grams of product by utilizing the formula weight.
- Smile, because you're done!!

Practice this process by completing the practice problems on the following pages.

Name_____ Date_____
Friendly Chemistry

Lesson 27: Stoichiometry Problems –1
Beginning with Milliliters of Solutions

Read each problem below. Work slowly and carefully.

1. Jenny had 45 mls of a 2 M sodium hydroxide solution. If she mixed all 45 mls of this solution with hydrogen chloride, how many grams of sodium chloride could she produce from this reaction. Water is a byproduct of the reaction.

2. Tony knew that ammonia (NH_3) will react with hydrogen sulfide to produce ammonium sulfide. If Tony had 100 mls of a 0.5 M ammonia solution, how many grams of ammonium sulfide could he potentially produce from this reaction? Assume Tony used all 100 mls of ammonia.

3. Julie had 50 mls of 5 M copper (II) sulfate. If she combined this compound with iron filings (shavings of iron metal), she could produce iron (II) sulfate and copper. How many grams of copper cold she potentially produce is she used all 50 mls of the 5M copper (II) sulfate solution?

4. Leroy took 1 liter of 0.25 M potassium chlorate and allowed it to decompose into potassium chloride and oxygen gas. How many grams of potassium chloride could you predict he generated from this reaction?

Name_____ Date_____

Friendly Chemistry

Lesson 27: Stoichiometry Problems –2
Beginning with Milliliters of Solutions

Read each problem below. Work slowly and carefully.

1. When zinc coated bolts are dropped into sulfuric acid (hydrogen sulfate), zinc sulfate and hydrogen gas are produced. Suppose Tim dropped some of these bolts into 200 mls of 2.0 M sulfuric acid. How many grams of zinc sulfate did he produce from this reaction?

2. Potassium perchlorate decomposes into potassium chloride and oxygen gas. If you begin with 50 mls of a 12 M potassium perchlorate solution, how many grams of potassium chloride might you produce from this reaction?

3. 400 mls of a 1.5 M iron (III) nitrate solution were mixed with lithium hydroxide. Iron (III) hydroxide and lithium nitrate were produced. How many grams of each product were made?

4. Tony had 5 moles of sodium acetate. Mary had 5 moles of barium acetate. Who had the most atoms? Show your work.

Lesson 28: Determining the Needed Amount of a Solution to Perform a Reaction

In Lesson 27 you learned how to take a known amount of solution of a known molarity and place it into a chemical reaction in order to predict the quantity of a product from the reaction. You learned that you first have to convert the milliliters of solution to grams of solute and then ultimately moles of solute before you could continue with the stoichiometry calculations. In this lesson you will learn how to determine the amount of known molarity solution necessary in order to produce a desired amount of product. This procedure is very similar to determining the amount of reactant necessary to produce a desired amount of product (Lesson 25). However, in this case instead of determining the amount of dry reactant needed, you will be asked how many milliliters of a known molarity solution it will take to make the desired amount of product.

Let's begin by looking at an example:

Friendly Chemistry

Example 1: Suppose that you've been asked to prepare 100 grams of potassium chloride for your boss. You've found in your store room that you have 5 M potassium chlorate which will readily decompose to produce potassium chloride and oxygen gas. How many milliliters of 5 M potassium chlorate will be necessary to make the 100 grams of potassium chloride?

Start by writing and balancing the chemical reaction.

$$K^{+1}_1(ClO_3)_1 \longrightarrow K_1Cl_1 + O_2$$

$$2K^{+1}_1(ClO_3)_1 \longrightarrow 2K_1Cl_1 + 3O_2$$

Identify the reaction components we are interested in.

$$\boxed{2K^{+1}_1(ClO_3)_1} \longrightarrow \boxed{2K_1Cl_1} + 3O_2$$

Then, determine the mole ratio.

2:2 which is the same as 1:1

Next, let's look back at our problem. How many grams of potassium chloride are we supposed to make? _____

Recall from Lesson 22 that we can only apply the mole ratio to moles of product or reactant. So, the next thing we'll need to do is convert these desired 100 grams of potassium chloride into moles of potassium chloride. And, since you've done this many, many times, you know you'll need a formula weight for the potassium chloride.

$$K_1Cl_1$$

K: 1 x 39 g = 39 grams
Cl: 1 x 35 g = <u>35 grams</u>
Formula weight = 74 grams

Now, we've been asked to make 100 grams of potassium chloride from this reaction. Knowing one mole of potassium chloride has a mass of 74 grams allows us to create this set of equivalent fractions:

$$\frac{1 \text{ mole KCl}}{74 \text{ grams}} = \frac{x \text{ moles}}{100 \text{ grams}}$$

$$x = 1.35 \text{ moles KCl}$$

Therefore based upon these calculations, we see we'll need to produce 1.35 moles of potassium chloride (KCl). Since we have this amount now in moles, we can apply the mole ratio we found earlier.

$$1 : 1$$

So, based upon this ratio, we know in order to produce 1.35 moles of potassium chloride, it will require 1.35 moles of potassium chlorate.

Now, check back to our problem. We are being asked to determine how many milliliters of a 5 M potassium chlorate solution are necessary to make the desired 100 grams of potassium chloride. If our available solution is 5 M, that means that there are 5 moles of potassium chlorate per liter (1000 mls) of solution. Knowing that, we can set up these two equivalent fractions:

$$\frac{\text{In 1000 milliliters solution}}{\text{There are 5 moles of K(ClO}_3\text{)}} = \frac{\text{How many (x) milliliters}}{\text{in 1.35 moles of K(ClO}_3\text{)}}$$

Or

$$\frac{1000 \text{ ml}}{5 \text{ m K(ClO}_3\text{)}} = \frac{x \text{ ml}}{1.35 \text{ m K(ClO}_3\text{)}}$$

$$5x = 1350$$
$$x = 270 \text{ ml K(ClO}_3\text{)}$$

Therefore, based upon these calculations, we can say that in order to prepare 100 grams of potassium chloride, we would need at least 270 mls of 5 M potassium chlorate.

Let's look at another example:

Example 2: If bromine gas is bubbled through a solution of potassium iodide, potassium bromide and iodine crystals are produced. Suppose that you've been asked to prepare 50 grams of iodine crystals (I_2) from this reaction. If the potassium iodide solution is 2.0 M, how many milliliters of this solution would you need?

Begin by writing and balancing the chemical reaction:

$$Br_2 + 2KI \longrightarrow 2KBr + I_2$$

Locate the parts of the reaction that you are concerned with:

$$Br_2 + \boxed{2KI} \longrightarrow 2KBr + \boxed{I_2}$$

And determine the mole ratio:

$$2 : 1$$

Look back at the problem now. We are being asked to prepare 50 grams of iodine crystals. Before we can apply the mole ratio to this desired amount of product, we must first convert it into moles. To do this we need to determine how many grams are in 1 mole of iodine crystals aka finding the formula weight of I_2.

$$I_2$$

I: 2 x 127 g = 254 grams / mole

Knowing this amount, we can then set up equivalent fractions to find out how many moles are equal to the 50 grams we are being asked to prepare.

$$\frac{1 \text{ mole } I_2}{254 \text{ g}} = \frac{x \text{ moles}}{50 \text{ g}}$$

$$x = 0.20 \text{ moles}$$

Based upon the calculations, we see that our 50 grams of iodine crystals are equivalent to 0.20 moles of iodine crystals. Now that we have this amount in moles, we can apply it to the mole ratio:

$$2 : 1$$

$$0.40 \text{ moles KI} \longrightarrow 0.20 \text{ moles I}_2$$

In order to make the desired 0.20 moles of iodine crystals, it will take twice as many moles of potassium iodide or 0.40 moles.

Recall that our potassium iodide is found in a 2 M solution. We know we will need 0.40 moles, but now need to know how many milliliters that is equivalent to. A 2M solution means that there are 2 moles of KI found in each liter (1000 mls). Knowing this we can set up two equivalent fractions:

$$\frac{\text{In 1000 mls there are}}{2 \text{ moles KI}} = \frac{\text{How many mls (x)}}{0.40 \text{ mols KI}}$$

$$2x = 400$$
$$x = 200 \text{ mls}$$

Based upon these calculations, we can say that it will take 200 mls of 2 M KI in order to make the desired amount of 50 grams of iodine crystals.

To review what you've learned in these examples, first we attacked the problem just like any other stoichiometry problem. We wrote and balanced the chemical reaction. We determined the mole ratio and used it to predict the amount of ingredient necessary to produce the desired amount of product. We then looked at the molarity of the reactant solution we had available and finally determined the volume necessary to meet our needs in the reaction.

Practice the techniques you've learned here on the following pages.

Name_____ Date_____
Friendly Chemistry

Lesson 28: Molarity Problems –1
Determining Volume of Reactant Necessary to Produce Desired Product

Read each problem below. Work slowly and carefully. Ask if you have questions.

1. Joe has been asked to prepare 300 grams of hydrogen cyanide for his boss. He is going to use the following reaction: Hydrogen sulfate and sodium cyanide react to yield hydrogen cyanide and sodium sulfate. He has 3 M hydrogen sulfate on hand for the reaction. How many milliliters of this solution will be required to make the 300 grams of H(CN)?

2. Patricia was preparing to make 450 grams of zinc chloride. She knew that when zinc oxide and hydrochloric acid are combined, they product zinc chloride and water. She looked on her supply shelf and found she had 5 M hydrochloric acid and plenty of zinc oxide. How many milliliters of the 5 M hydrochloric acid will she need?

3. Terry knew that carbon and aluminum oxide will react to produce aluminum and carbon dioxide. If Terry had a container of 5 M aluminum oxide in his supply cupboard, how many milliliters would it take for him to prepare 500 grams of aluminum based upon his reaction?

4. Martha read that iron (III) nitrate and lithium hydroxide would undergo a double replacement reaction to produce iron (III) hydroxide and lithium nitrate. If Martha took 100 mls of a 3.5 M lithium hydroxide solution, how many grams of lithium nitrate could she expect to produce?

Name_____ Date_____
Friendly Chemistry

Lesson 28: Molarity Problems –2
Determining Volume of Reactant Necessary to Produce Desired Product

Read each problem below. Work slowly and carefully.

1. When zinc is placed into sulfuric acid (hydrogen sulfate), zinc sulfate and hydrogen gas are produced. Suppose you needed to produce 1000 grams of zinc sulfate from this reaction. You had plenty of zinc on hand and found you had 750 milliliters of 6 M sulfuric acid available for use. Would you have enough? Show your work as proof of your answer.

2. Potassium perchlorate decomposes into potassium chloride and oxygen gas. If you have instructions of prepare 1 kilogram of potassium chloride for a lab activity, how many milliliters of a 0.5 M potassium perchlorate solution would you need?

3. Trey needed to prepare 5 kg of sodium chloride from the reaction of hydrogen chloride and sodium hydroxide. If his hydrogen chloride solution was 2 M, how may milliliters would he need? Water is a byproduct of the reaction.

4. Melissa had 500 ml of a 2 M barium oxide solution. If she bubbled carbon dioxide through this solution, how many grams of barium carbonate could she expect to produce?

Lesson 29: Properties of Gases and How Gases can be Measured in Reactions

In many of the last lessons, we have looked at reactions where gases may be used as a reactant in the reaction or be a product created by the reaction. We tended to treat them like a solid or liquid in that we said we were using a certain number of moles or grams of gas as a reactant or we had produced a certain number of moles or grams of gas from the reaction. In reality, measuring amounts of gases can be rather tricky. This trickiness comes from the unique properties or characteristics of gases. In this lesson we will take a closer look at the properties of gases and how we can measure them in more meaningful ways.

Before we zero in on gases, let's take a minute to look at solids and liquids. Suppose you have a block of wood. You would probably classify that block of wood as being a solid in that it has its own definite shape and a definite volume. If you put that block of wood into a round jar, it still keeps its block-like shape. Solids have a definite shape. You could measure its volume (the amount of space it takes up) by measuring a length, width and height and multiplying those together to arrive at a volume for the block of wood. You could also find its volume by lowering into a container of water and see how much water gets displaced by the block. Solids have a definite shape and volume.

Let's consider liquids next. Suppose you pour some milk into a drinking glass. Unlike the solid block of wood, the milk will not keep its own shape. Instead, it takes the shape of the inside of the glass. No matter where we put the milk, it always takes the shape of the object in which it is placed. This is the same for all liquids. We can say, then, that liquids, unlike solids, don't have a definite shape.

However, like solids, liquids do have a definite volume. Think back to the glass of milk. When we put the milk into the glass, we could measure the amount of milk we put in using units like ounces or milliliters. If we poured this same milk into another cup or glass, we could measure its volume again and would most probably find the volume to be the same (or very close to the same as a few drops may have stayed behind in the first glass). No matter which container we put the milk into, the milk will always have its own volume. So, like solids, liquids have a definite volume that can be measured.

Let's recap here: so far, we've said that solids have their own shape and definite volume. Liquids also have their own volume, but their shape takes on the shape of whatever we've put them into. Now, let's look at gases.

Think about blowing up a balloon. As you blow up the balloon, you are forcing air into the balloon. Does the gas go to all parts inside the balloon or do we just find air filling up the lower part of the balloon like a liquid might? It's easy to see that the air goes everywhere inside the balloon. Once the balloon is filled and the opening tied shut, no matter how you squeeze the balloon, the air fills up all available space inside the balloon.

Suppose you have a container nearby into which you could transfer the air from the inflated balloon. Suppose this container was completely empty when you began the transfer and was sealed to the air around it. As you transferred the gas from the balloon into this new container, does it go just to the bottom of the container? Of course not; it again fills up whatever space is available in the new container. No matter where we put this air, it always tries to fill up the entire space where we put it. Gases fill up the entire container in which they are being held—they do not have a definite volume. This makes it really tricky when we try to measure volumes of gases.

In addition, because gases fill up the entire space inside the container in which they have been placed, they always take the shape of the inside of that container. Gases do not have a definite shape or volume!

To review what we've discussed so far, we've said that solids have definite shapes and volumes. Liquids have definite volumes but take the shape of the container in which they have been placed. Gases do not have definite shape or definite volumes!

To explain why this variation in properties of solids, liquids and gases might exist, we need to think about the atoms which make up matter. Chemists and physicists have theorized that matter that is a solid has its atoms lined up in relatively straight patterns and the atoms sort of just vibrate back in forth in place. The atoms in liquids, on the other hand, have a greater amount of energy and instead of vibrating back and forth like in solids, the atoms of liquids slide over each other creating matter that doesn't hold its own shape. The theory goes on to say that the atoms of gases have much greater amounts of energy than both solids and liquids. In fact, the atoms of gases have so much energy that the atoms fly about filling up the entire space that they occupy. It is this great amount of energy present in atoms of gases which causes them to not have a definite volume nor shape.

Because gases occupy the entire container in which they are in, the atoms making up those gases find themselves "banging" into the sides of the container quite often. This collective "banging" is what creates gas pressure. The more "banging" going on, the greater the pressure. In one's car tires, there's an incredible amount of "banging" going on, hence high air pressure. There are factors which increase gas pressures: temperature and volume.

Let's turn back to one of our original objectives of this lesson: how we measure gases in reactions. Because volumes of gases vary according to temperature and pressure, we need to somehow standardize the means of measuring gases. This standardization is done by measuring gases at specific conditions known as STP. **STP stands for Standard Temperature and Pressure**. If we all agree to measure gases at STP (or at least convert the amounts we're using to volumes at STP) then we can communicate and fairly trade and market gases with one another.

So, you may be asking how does this all tie into measuring gases produced by chemical reactions? There's one additional piece of information necessary to bring this into focus. Recall the famous chemist we discussed several lessons ago by the name of Avogadro (remember Avogadro's number?). It so happens that Dr. Avogadro also discovered the idea that a mole of a gas is proportional to the volume of gas present. In other words, if you have one mole of a gas at a constant temperature and pressure, like STP, it will have a set volume. If you double the moles of that same gas, keeping it at STP, you will double the volume. If you triple of number of moles present, you triple the volume remembering all the while, that you are keeping the temperature and pressure constant. There is a direct relationship between moles of a gas present and volume of that gas. This relationship is known as Avogadro's law.

This can be carried one step further. If a gas is behaving like most gases do, it can be said to be behaving like an **ideal gas**. One mole of an ideal gas has a volume of 22.4 liters. Two moles would have volume of 44.8 liters, etc. Another way to look at this relationship is 22.4 liters of an ideal gas contains 1 mole of that gas. This concept is so important that we need to say it again: One mole of an ideal gas at STP has a volume of 22.4 liters. Or, if you have 22.4 liters of a gas at STP, you have 1 mole of that gas. Let that sink in and when ready, read on.

Example 1: Let's look at an example of a stoichiometry problem you worked with earlier. Suppose you had 44.8 liters of fluorine gas which you exposed to some potassium fluoride. How many moles of chlorine gas could you produce from this reaction? And then, how many liters of chlorine gas would this be equivalent to? Potassium fluoride is a byproduct of this reaction. Assume you are working at STP.

Begin by writing and balancing the chemical equation.

$$F_2 + 2KCl \longrightarrow Cl_2 + 2KF$$

Balances to become...

$$F_2 + 2KCl \longrightarrow Cl_2 + 2KF$$

Like before, identify the components within the reaction that we are interested in. In this problem we are given 44.8 liters of fluorine gas and are being asked to determine the number of moles of chlorine gas produced by the reaction.

$$\boxed{F_2} + 2KCl \longrightarrow \boxed{Cl_2} + 2KF$$

Continue by identifying the mole ratio. In this problem there is a 1:1 ratio (1 mole of Cl_2 is produced for every 1 mole of F_2 placed into the reaction).

$$1 : 1$$

Look back at the problem. We have been told we have 44.8 liters of chlorine gas. Based upon what you've learned so far in this lesson, you know that 22.4 liters of an ideal gas contains one mole of that gas. In this case we have 44.8 liters. We can make this relationship:

$$\frac{1 \text{ mole gas}}{22.4 \text{ liters}} = \frac{x \text{ moles of gas}}{44.8 \text{ liters}}$$

Solving for x, we find that our 44.8 liters of fluorine gas are equivalent to 2 moles of fluorine gas. So, we're beginning our reaction with 2 moles of fluorine gas.

As before, we can now apply the mole ratio. We determined above that in this reaction it was a 1:1 ratio, therefore, if we put in 2 moles of fluorine gas, we should produce 2 moles of chlorine gas.

Finally, in the problem, we were asked to determine how many liters of gas we would produce. Knowing that 1 mole of an idea gas at STP occupies 22.4 liters of space, the two moles of chlorine we produced in this reaction would then be equivalent to 44.8 liters of chlorine gas. Let's look at one more problem.

Example 2: Suppose you have been given a 45 gram sample of hydrogen peroxide. You exposed this sample to sunlight and it decomposed to water and oxygen gas. If all of the hydrogen peroxide decomposed, how many liters of oxygen gas could you produce from this reaction. Assume you are working at STP.

Begin by writing and balancing the reaction:

Friendly Chemistry

$$H_2O_2 \longrightarrow H_2O + O_2$$

Becomes the balanced reaction:

$$2H_2O_2 \longrightarrow 2H_2O + O_2$$

Locate the components in the reaction in which we are interested.

$$\boxed{2H_2O_2} \longrightarrow 2H_2O + \boxed{O_2}$$

Determine the mole ratio. In this case the mole ratio is:

$$2 : 1$$

Now, look back at the problem. How much hydrogen peroxide are we starting with? _____ (45 grams, right?). Before we can apply the mole ratio to this amount, we must first do what? _____
(hopefully you said, "Convert it to moles!") In order to do that we must find the formula weight of the hydrogen peroxide. Do that here:

$$H_2O_2$$

H:

O:

Formula wt. = _____

So, our 45 grams of H_2O_2 is equivalent to _____ moles (45 g / 34 g/mole)

Now, we can apply the mole ratio to the amount of moles of H_2O_2 we are beginning with (1.32 moles). Since we have a 2:1 ratio, we'll be producing half the number of moles of O_2 compared to the number of moles of H_2O_2 that go into the reaction.

$$2 : 1$$
$$1.32 \text{ moles of } H_2O_2 \longrightarrow 0.66 \text{ moles of } O_2$$

From these calculations we now know from the 45 grams of hydrogen peroxide that we began with, we found we will be producing 0.66 moles of oxygen gas. Based upon what we've learned in this lesson, we know that 1 mole of an ideal gas has a volume of 22.4 liters while at STP. Knowing this, we can setup this relationship:

$$\frac{1 \text{ mole gas}}{22.4 \text{ liters}} = \frac{0.66 \text{ moles of } O_2}{x \text{ liters}}$$

And solving for x, we get:

$$x = 14.78 \text{ liters of } O_2$$

After this last calculation we can now say that from our 45 grams of hydrogen peroxide with which we began, we will produce 14.78 liters of oxygen gas at STP.

Let's stop and review what we've discussed in this lesson. We said that:
- Matter that is a solid has a definite shape and volume.
- Matter that is a liquid has a definite volume but takes the shape of container that it occupies (no definite shape).
- Matter that is a gas has no definite volume and completely fills the container it occupies. Gases have no definite volume or shape.
- The volume of a gas is measured at STP (Standard Temperature and Pressure).
- One mole of an ideal gas has the volume of 22.4 liters at STP.
- 22.4 liters of an ideal gas is equivalent to one mole of that gas at STP.

Practice what you've learned in this lesson by completing the practice problems on the following pages..

Friendly Chemistry

Name_____ Date_____
Friendly Chemistry

Lesson 29: Measuring Molar Volumes of Gases

1. You have learned that one mole of any gas at STP, occupies a volume of 22.4 liters. If Gina has 44.8 liters of a gas, how many moles of that gas does she have?

2. Georgette collected 18 liters of neon gas. Does she have less than a mole, exactly a mole or more than a mole of neon?

3. Carbon dioxide gas is a product from the reaction of acetic acid (vinegar) and baking soda. Suppose you completed this reaction and found that you collected 50 liters of carbon dioxide at STP. How many grams of carbon dioxide did you produce in this reaction?

4. Tony had 12 moles of nitrogen gas (N_2). How many grams is this equal to?

5. Argon gas is used between the glass panes of insulating windows. If the space between two panes of glass was equal to 3.5 liters, how many moles of argon gas could you inject at STP?

6. Potassium chloride and fluorine gas react to produce potassium fluoride and chlorine gas. If Patty begins with 24 moles of potassium chloride, how many liters of chlorine gas might she expect to produce in this reaction?

7. When a person mixes calcium with water, calcium hydroxide and hydrogen gas are produced. If you begin with 400 grams of calcium, how many liters of hydrogen gas can you expect to produce from this reaction?

8. Tony had 90 milliliters of water. If he applied electricity to this water it would decompose to produce hydrogen gas and oxygen gas. How many liters of oxygen gas might he expect to produce from this 90 mls of water. He knew that one ml of water had the mass of 1 gram.

9. Potassium chlorate will decompose to produce potassium chloride and oxygen gas. If Hank began with 45 kilograms of potassium chlorate, how many liters of oxygen gas measured at STP might he expect to produce from his reaction?

10. How many atoms might you expect to find in 134.4 liters of helium gas?

Friendly Chemistry

Lesson 30: How Changing Temperatures affect Gases

In the last lesson you learned that you could measure volumes of gases either used as a reactant or produced in a chemical reaction provided you kept the gas at STP (standard temperature and pressure). However, gases are seldom stored at STP. It makes sense to understand how changing the temperature or pressure placed upon a gas affects its volume. In this lesson, we will focus on how temperature changes affect the volume of a gas. In Lesson 31 we will look at how changing the pressure of a gas affects its volume.

An easy to visualize scenario of how temperature affects a gas is to consider how a hot air balloon works. At the base of the balloon there is a large burner or heat source. When the balloonist is ready to launch his balloon, he "fires-up" the burner and begins to heat the air inside the balloon. As the air warms, the balloon begins to expand and eventually lifts the balloonist and his basket off of the ground. On an atomic level, the atoms of the gases present in the air gain more and more energy and find themselves moving faster and faster. Because they are moving faster, more and more of them find themselves hitting the sides of the balloon more often. Because of all this increase in activity, we can say that as the air heats, the volume of the gas increases. In other words, as the temperature of the gas increases, so does the volume. The opposite also happens as a gas is cooled. When the hot air balloonist is ready to descend to the ground, he will lower or even turn off his burner. Note that this relationship is a direct relationship—as one increases so does the other and vice versa, as one goes down, so does the other. One must realize that in this balloon illustration, that the pressure outside the balloon (atmospheric pressure in this case), remained the same.

The scientist who developed this principle was a French scientist named Jacques Charles. In fact, Charles was the first person to ever fill a balloon with hydrogen gas and make the first solo flight! The relationship between temperature and volume of a gas became known as Charles's Law.

Let's look now at some applications of this concept to laboratory situations.

Example 1: Suppose you had 5 liters of hydrogen gas. The temperature of the gas was room temperature (25 degrees C.). In order to use this gas in stoichiometric problems, we must determine its actual volume at STP.

Recall that STP stands for Standard Temperature and Pressure. Standard Temperature is exactly 0 degrees Celsius. From our discussion above, we noted that as the temperature falls, so does the volume of a gas. In this case, our gas was originally at 25 degrees C. Since Standard Temperature is less than 25 degrees C, we can guess that the "new" volume will be less than the original volume. However, before we can apply Charles' Law to this problem, we must first convert these Celsius temperatures to Kelvin temperatures.

In order to convert Celsius temperatures to Kelvin, we simply add 273 degrees to the Celsius temperature. Zero degrees C + 273 = 273 degrees K. Therefore, we can say that Standard Temperature is 0 degrees C or 273 K.

Looking back at our example above, we'll also need to convert the 25 degrees C that are gas was originally at to Kelvin degrees. Again, we simply add 273 degrees to the 25 = 298 degrees K.

If one were to take a substance clear down to zero degrees K (not C, now), this point is called absolute zero. It is theorized that at this point, all movement within the atoms making up that substance cease. Now back to our example:

Based upon Charles's Law, we can set up the following relationship:

$$\frac{\text{Original Temperature}}{\text{New Temperature}} = \frac{\text{Original Volume}}{\text{New Volume}}$$

Or

$$\frac{T_1}{T_2} = \frac{V_1}{V_2}$$

Where T_1 and V_1 = original temperature and volume and T_2 and V_2 = new temperature and new volume.

So, if we substitute in the values from our example:

$$\frac{298 \text{ K}}{273 \text{ K}} = \frac{5 \text{ L}}{x \text{ L}}$$

Solving for x, we get $298x = 273(5)$

$$298x = 1365$$

$$x = 4.58 \text{ L}$$

Therefore, we can say that our given 5 L of gas at room temperature is actually only 4.58 L when we adjust its volume to STP. As we predicted, the volume of the gas decreased as the temperature fell. Now, we can utilize this "actual" volume of gas present in our stoichiometric problems.

Friendly Chemistry

Example 2: Suppose you had 17 L of hydrogen gas at STP. Inside your lab the temperature is 28 degrees C. What would the new volume of hydrogen gas be now in your lab?

First, let's look to see how the temperature of the gas changed. Did the temperature go up or down? _____

Yes, the temperature did rise. If the temperature rises, how do we expect the volume to change? _____. Let's see if this is the case by continuing the problem.

Fill in these blanks:

$T_1 =$ _____ C + 273 = _____ K

$T_2 =$ _____ C + 273 = _____ K

Now, the original volume (V) = _____ L

Let's put these values into a relationship according to Charles' Law:

$$\frac{T_1}{T_2} = \frac{V_1}{V_2}$$

So, substituting our values from our example:

$$\frac{273 \text{ K}}{301 \text{ K}} = \frac{17 \text{ L}}{x \text{ L}}$$

Solving for x, we get:

$$273X = 301(17)$$
$$273x = 5117$$
$$x = 18.7 \text{ L}$$

Did the volume increase as we expected? _____. We can say that due to an increase in temperature, the volume of our hydrogen gas increased from 17 L to 18.7 L. This concurs with our prediction that the volume would rise.

To review what we've learned here, we said that the volume of a gas is dependent upon its temperature. As the temperature of a gas goes up, the atoms making up that gas have more energy and therefore the volume of the gas also goes up. The opposite is true in that as a gas is cooled, its volume decreases. We learned this relationship was developed by a scientist named Charles. Using Charles's Law, we could write a mathematical relationship between the original temperature and new temperature and the original and "new" volume. Finally, we learned that while Standard Temperature was 0 degrees C, we needed to convert temperatures given in Celsius to the Kelvin (K) scale before we used Charles' Law.

Practice your gas volume prediction skills by completing the following practice pages.

Friendly Chemistry

Name_____ Date_____

Friendly Chemistry

Lesson 30: Charles's Law Practice –1
How Temperature Affects the Volume of a Gas

1. Tory had 5 L of helium gas at STP. If she heated the helium to 30 degrees C, what would the new volume of gas be?

2. Amelia had 4.5 L of oxygen gas which was being held at 10 degrees C. If the gas was heated by 20 degrees C, what would the new volume of gas be?

3. Tim and Margaret each had a gas-filled balloon. The volumes of their balloons were exactly 2 L at STP. If Tim heated his balloon by 30 degrees K, what would the new volume of his balloon become?

4. Jack filled a balloon with 4 liter of carbon dioxide gas. The gas was at room temperature (25 degrees C). If he took it outdoors where the temperature was 4 degrees C, what would the new volume of the balloon be? Assume the pressure outside the house was the same inside the house.

5. Sheila worked in a lab which collected neon gas. Suppose she collected 45 liters of neon at 35 degrees C. If she cooled this gas to STP, what would be the new volume of the gas?

6. Freddy purchased 14 L of argon gas at 28 degrees C. If he expected to store the gas at STP, how many L would he expect to need room for?

7. Mary Jo had a balloon filled with 0.5 liters of helium. If the temperature of the gas rose by 15 degrees, what would the new volume of her gas-filled balloon become?

8. Morris collected some hydrogen gas from an experiment he had conducted. He collected 12 L at a temperature of 28 degrees C. How many liters would this be equal to at STP?

Name_____Date_____
Friendly Chemistry

Lesson 30: Charles's Law Practice –2
How Temperature Affects the Volume of a Gas

1. Calcium fluoride and sulfuric acid react to produce calcium sulfate and hydrogen fluoride gas. If you begin with 13 moles of calcium fluoride, how many liters of hydrogen fluoride gas measured at STP should you produce from the reaction?

2. If the HF gas you produced in question 1 was heated <u>by</u> 40 degrees C, what would the new volume of HF be?

3. When strontium is placed into nitric acid (hydrogen nitrate), strontium nitrate and hydrogen gas are produced. If you begin with 58 grams of strontium, how many liters of hydrogen gas can you produce at STP?

4. If the H_2 gas you produce in question 3 is heated <u>to</u> 25 degrees C, what will the new volume of gas become?

5. Iron (II) sulfide and hydrochloric acid react to produce iron (II) chloride and hydrogen sulfide gas. Suppose you begin with 1.5 kilograms of iron (II) sulfide, how many liters of hydrogen sulfide gas could you produce from this reaction assuming you conduct the reaction at STP?

6. Suppose you have 48 liters of this hydrogen sulfide gas at room temperature (25 degrees C.) If you cool this gas to STP, what would the new volume be?

Lesson 31: How Changing Pressures affect Gases

In Lesson 30 you learned how volumes of gases are affected by temperature changes. You learned that as the temperature of a gas went up, the volume went up and as the temperature went down, the volume would also go down. This was a direct relationship. In this lesson, we will examine what happens to a gas when pressures placed upon the gas change.

The Irish scientist, Robert Boyle, made several observations as to how gases behave when pressures placed upon those gases changes. He noted that if you took a trapped sample of gas and increased the pressure placed upon it, the volume of the gas would decrease. He also noted that if you decrease the pressure placed upon a trapped gas, the volume would increase.

This relationship between pressure and volume became known as Boyle's Law. Note that this relationship, unlike Charles's Law, is an **inverse relationship**: as one variable goes up, the other comes down and vice-versa. Note that this relationship holds true as long as the temperature of the gas remains constant.

Knowing this information, we can predict then how a sample of gas will respond if the pressure being applied to in changes. Like you learned with Charles's Law, we can use a series of relationships to help us predict what might happen to the volume of gas should we change the pressure. Let's look at an example.

Example 1: Suppose you have 6 liters of neon gas at 635 mm of Hg. Predict how the volume of neon will change should you increase the volume to 785 mm of Hg. Assume the temperature does not change.

Before we take on this problem, let's take a minute and look at the units used to measure pressure. In this problem the units of pressure were mm of Hg or millimeters of mercury. When Boyle was making his observations of how gases behaved he used a J-shaped tube with a sample of gas trapped inside. To the open end he poured in mercury. He measured the height of the mercury column and compared it to the volume of gas trapped in the tube. While Boyle used inches of mercury, scientists today use the metric system hence millimeters of mercury to indicate degree of pressure present. Millimeters of mercury is often called the torr in honor of Evangelista Torricelli, an Italian scientist who built the first barometer.

A related unit for pressure is the standard atmosphere (atm for short). The relationship between mm of Hg and torr and atm is this:

$$1 \text{ atm} = 760 \text{ mm Hg} = 760 \text{ torr}$$

There is another unit of pressure which is the standard international unit (SI unit) known as the pascal or Pa for short.

$$1 \text{ atm} = 101,325 \text{ Pa}$$

This means that the pascal is a very small unit and because of this, it won't be used in any problems you encounter in this lesson.

Finally, there is another unit of pressure that you may be more familiar with which is called psi which stands for pounds per square inch. This unit is used quite often in engineering sciences. You may recognize it as units used when filling air into a car tire. One atmosphere is equal to 14.69 psi.

So, in regard to units of pressure, mm of Hg will be the units used in the problems you are asked to solve in this lesson.

Let's return to our example problem now. We'll state it again here: Suppose you have 6 liters of neon gas at 635 mm of Hg. Predict how the volume of neon will change should you increase the volume to 785 mm of Hg.

Like we did with problems using Charles's Law, first we need to check and see what is the change in conditions which was stated in the problem. In this case, the pressure being applied to the gas changes from 635 mm Hg to 785 mm Hg. The pressure has gone up. According to Boyle's law, as the pressure applied to a trapped gas increases, the volume goes down (an inverse relationship). Therefore, we should find that the solution to this problem to be a volume less that what we started with which was 6 liters.

Let's identify some components in the problem first. The initial volume (V) of the neon gas is 6 liters. The initial pressure (P) of the neon was 635 mm Hg. The new pressure (P') is 785 mm Hg. We are being asked to find the new volume (V').

We can set up this relationship:

$$\frac{P_1}{P_2} = \frac{V_2}{V_1}$$

So substituting in the values in the problem, we get:

$$\frac{635 \text{ mm Hg}}{785 \text{ mm Hg}} = \frac{V_2}{6 \text{ L}}$$

V' = 635mm Hg / 785mm Hg X 6 L

V' = 4.85 L

Based upon our calculations, our new volume of neon will be 4.85 L. This value is indeed less than the 6 L of neon that went into the new conditions of an increase in pressure.

Let's look at a second example.

Example 2: Suppose you have a balloon holding 4.5 liters of oxygen gas at 735 mm of Hg. If the pressure falls to 710 mm of Hg, what will the volume of the balloon be changed? Begin by asking yourself what has changed in the conditions. In this case, the pressure has <u>decreased</u> from 735 mm Hg down to 710 mm Hg. Because this is a Boyle's law situation, we know that this will be an inverse relationship. So as pressure falls, volume should go up. Our answer, then, should be greater than 4.5 liters. Let's work it out and see!

Identify your components:

$$P_1 = 735 \text{ mm Hg}$$
$$P_2 = 710 \text{ mm Hg}$$
$$V_1 = 4.5 \text{ L}$$
$$V_2 = ?$$

Set up the mathematical relationship:

$$\frac{P_1}{P_2} = \frac{V_2}{V_1}$$

So substituting in the values in the problem, we get:

$$\frac{735 \text{ mm Hg}}{710 \text{ mm Hg}} = \frac{V_2}{4.5 \text{ L}}$$

$$V_2 = 735 \text{ mm Hg} / 710 \text{ mm Hg} \times 4.5 \text{ L}$$

$$V_2 = 4.66 \text{ L}$$

Our calculations say our new volume of gas will be 4.66 L which is greater than the 4.5 L that we began with. Since the pressure fell, the volume had to go up!

Before you move on to more practice problems, let's review what we learned in this lesson. You learned that when pressures are applied to gases their volumes will change. This relationship is an inverse relationship in that as pressure is increased, volume will be decreased and vice versa. We learned that there are several units used to measure pressure and finally, that we could use the known relationship between pressure and volume to predict volume changes due to changes in pressure.

Friendly Chemistry

Name_____ Date_____

Friendly Chemistry

Lesson 31: Boyle's Law Practice
How Pressure Affects the Volume of a Gas

1. Terry has 17 liters of helium gas at 760 mm Hg. If he increases the pressure to 790 mm of Hg, what will the new volume of helium be? Assume there is no temperature change.

2. The current atmospheric conditions were 765 mm Hg. Marty had 5 liters of hydrogen gas. If the pressure changed to 740 mm of Hg, what would the new volume of hydrogen be? Assume there is no temperature change.

3. Darcy had a weather balloon with 10 liters of helium inside it. If she took it into the mountains where the pressure was 20 mm of Hg less than where she started, what will the new volume of helium be in her balloon. Assume there was no temperature change.

4. Josh had 34 liters of argon gas under a pressure of 770 mm of Hg. If the pressure changed to 790 mm of Hg, what would the new volume of argon be? Assume the temperature did not change.

5. Sarah collected 500 ml of hydrogen from the reaction of zinc and hydrochloric acid. If she collected this at 1 atm of pressure but then pressurized the sample to 700 mm of Hg, how much space would the sample now require? Assume the temperature did not change.

Lesson 32: Combining the Gas Laws

In the last two lessons you learned how the volume of a gas is affected both by temperature and pressures which are applied to the gas. You learned that the relationship between temperature and the volume of a gas is a direct relationship (as the temperature goes up, so goes the volume and vice versa). Then you learned that the relationship between pressure and volume is an inverse relationship (as the pressure goes up, the volume goes down and vice versa). You also learned that in order to accurately measure a gas that it had to be placed under special conditions called STP. Measuring a gas at STP allows you to communicate with others the actual volume of gas present. In Lessons 30 and 31 you used mathematical relationships to convert gas volumes to STP. In this lesson we will combine those two steps into one.

Let's begin by looking at the mathematical relationship we used in Lesson 30 where we learned how temperature affects the volume of a gas. We used fractions like these:

$$\frac{T_1}{T_2} = \frac{V_1}{V_2}$$

Where T_1 = the original temperature and T_2 = the new temperature and where V_1 = the original volume and V_2 = the new volume.

With a little algebraic transformation (don't panic, just a little rearrangement of the variables), we can get this:

$$T_1 V_2 = T_2 V_1$$

Then, with solving for V_2 (the new volume), we get:

$$V_2 = V_1 (T_2 / T_1)$$

This basically says that we can find the new volume by multiplying the original volume by the ratio of the new temperature to the old temperature.

If we take this expression and then add the relationship which tells how volume is affected with changes in pressure, we get this:

$$V_2 = V_1 (T_2 / T_1) (P_1 / P_2)$$

This combines both of the relationships into one expression which basically says we can find the new volume by taking the original volume and multiplying by the ratio of new temperature to original temperature and by the ratio of original pressure to new pressure.

This may be getting confusing to you by this point so let's look at an example to help sort things out.

Example 1: Suppose you have collected 5 liters of oxygen gas at room conditions of 740 torr and 27 degrees C. What would the volume of oxygen be at STP?

Friendly Chemistry

To keep things organized, locate the following pieces of information:

T_1 = 27 C

T_2 = 0 C

P_1 = 740 mm Hg

P_2 = 760 mm Hg

V_1 = 5 L

V_2 = this is what we're looking for, right?

Recall that before we can utilize the temperature measurements, we must first convert the Celsius values into Kelvin values. Do so by adding 273 to each value so our measurements now become:

T_1 = 27 + 273 = __300__ degrees K.

T_2 = 0 + 273 = __273__ degrees K.

P_1 = 740 mm Hg

P_2 = 760 mm Hg

V_1 = 5 L

V_2 = ?

Before we put these values into our combined gas law formula, take a look at what happens when we convert the room conditions to STP. What happens to the temperature? Does it go up or down? _____ Hopefully you said it goes down! If the temperature goes down, what happens to the volume of the gas? _____ Yes, since this is a direct relationship, the volume should go down.

Now, consider the pressure side of the situation. The room pressure was 740 torr. What happened when you go from room conditions to STP? _____ Hopefully you said that the pressure goes up! Recall that the relationship between volume and pressure is an inverse relationship. As pressure goes up, volume comes down. So, again we should see a decrease in volume. Let's substitute in our values from our example and see the result!

Here's the combined gas law formula again:

$$V_2 = V_1 \; (T_2/T_1) \; (P_1/P_2)$$

After substituting in our values, we get:

S303

$$V_2 = 5 \text{ L } (273/300)(740/760)$$

Continuing to solve.....

$$V_2 = 5 \text{ L } (0.91)(0.97)$$

And after another step we get:

$$V_2 = 5 \text{ L } (88.27)$$

$$V_2 = 4.41 \text{ L of oxygen gas}$$

So, yes our volume did go down as expected ($V_1 = 5$ L, $V_2 = 4.41$ L)

Let's try another example.

Example 2: Tony was filling the tires on his four-wheeler with air before he went riding up into the mountains. At the gas station where he was filling the tires, the ambient temperature (ambient temperature means surrounding temperature or local conditions) was 34 C and the pressure was 762 torr. When he got up into the mountains the temperature was now 30 C and the pressure had fallen to 755 torr. If the volume of air inside each of Tony's tires was 24 L at the gas station, what was the new volume of the tires once he reached the top of the mountain?

Again, start by identifying measurements in the problem:

$T_1 = 34 + 273 = $ _____ degrees K.

$T_2 = 30 + 273 = $ _____ degrees K.

$P_1 = 762$ mm Hg

$P_2 = 755$ mm Hg

$V_1 = 24$ L

$V_2 = $ this is what we're looking for!

In looking at this situation, we see the temperature of the gas went _____ which would make us think that the volume of the gas will also go _____. However, take a look at the pressure change. The pressure goes _____ and since pressure and volume have an inverse relationship, the volume should go _____. So, in this case we have two "competing" factors: one making the volume decrease and the other making the volume increase.

Let's substitute our values in the combined gas law formula and see what the final results will be:

$$V_2 = V_1 \ (T_2/T_1) \ (P_1/P_2)$$
$$V_2 = 24 \text{ L} \ (303/307) \ (762/755)$$
$$V_2 = 24 \text{ L} \ (0.986) \ (1.01)$$
$$V_2 = 24 \text{ L} \ (0.996)$$
$$V_2 = 23.9 \text{ L}$$

So, in this case there was a slight decrease in overall volume of the air inside Tony's four wheeler tires.

Practice this technique by working through the problems on the Practice Pages which follow.

Parting Comments: As you are aware, Lesson 32 is the final lesson in your *Friendly Chemistry* course. By completing this lesson, you will have accomplished a great deal of work and will have built a very solid foundation in chemistry. We wish you the very best in your future studies and welcome any feedback you can provide to help us improve this course. If we can assist you in the future, please let us know!

Dr. Joey and Lisa Hajda

Friendly Chemistry

Name_____ Date_____

Friendly Chemistry

Lesson 32: Combining the Gas Laws

1. Suppose you collected 10 L of hydrogen gas from a lab experiment when the room temperature was 34 C and the atmospheric pressure read 762 torr. What would the volume of this hydrogen be at STP?

2. Georgia had 3.5 L of argon gas at 33 C and a pressure of 733 torr. If she heated the gas to 40 C and decreased the pressure <u>by</u> 30 torr, what would the new volume of gas be?

3. Sebastian combined zinc metal and hydrochloric acid to produce zinc chloride and hydrogen gas. If he collected 7 liters of hydrogen from this reaction when the room temperature was 24 C and the pressure was 755 torr, what would the volume be at STP?

4. Tommy had 3.5 L of neon gas at 22 C. He heated it up to 32 C and increased the pressure to 780 torr. If the original pressure was 760 torr, what would the new volume of the neon be?

5. Molly collected 10 liters of fluorine gas from a lab procedure. The room in which she collected it was at a temperature of 26 C. If she placed it into a cooler where the temperature was 20 C, what would the new volume of gas be. Pressures inside and outside the cooler were the same.

6. How large would a 100 L sample of gas become if the pressure went from 760 torr down to 2 torr?

Friendly Chemistry

Index

Atom, S20
Atomic radius, S89
Avogadro's number S141
Balancing chemical equations, S201
Bingo, element, S6
Boyle's Law, S295
Charles' Law, S285
Chemical equations, balancing, S201
Chemistry, definition of, S2
Compound formation, S118
Electron dot notation, S61
Electronic configuration notation (ECN), S53
Elements, S3
Element families, S71
Empirical formulas, calculating, S165
Endothermic reactions, S187
Exothermic reactions, S187
Formula weights, calculating, S147
Gas, definition of, S228
Ion formation, S107
Ionization energy, S82
Law of Conservation of Matter, S201
Liquid, definition of, S228
Magnetic Quantum number, S32
Molar volume of a gas, S278
Molarity, S248
Mole, introduction activity, S139
Orbital Quantum number, S31
Orbital notation, S45
Polyatomic ions master list, S129
Percent composition, calculating, S157
Principle Quantum number, S29
Quantum mechanics, S29
Reactions, S183
Solid, definition of, S228
Solute, S247
Solutions, S247
Solvent, S247
Spin Quantum number, S34
Stoichiometry, S213
STP Standard Temperature and Pressure, S229
Subatomic Particles, S21
Valence electron, S61

Useful Conversions

1 pound = 454 grams
2000 pounds = 1 ton
1 quart = 0.946 liters
2 pints = 1 quart
4 quarts = 1 gallon
2 cups = 1 pint = 16 fluid ounces
1 cup = 8 ounces
3 teaspoons = 1 Tablespoon
1 cubic centimeter (cc) = 1 ml
1 inch = 2.54 cm
1 angstrom = 1×10^{-8} cm
1 micron = 1×10^{-6} meters
Mega = million
Kilo = thousand
Deca = ten
Deci = one-tenth
Centi = one-hundredth
Milli = one-thousandth

Temperature Conversions

Celsius to Fahrenheit: 9/5 (C) + 32
Fahrenheit to Celsius: 5/9 (F) - 32
Celsius to Kelvin: C + 273

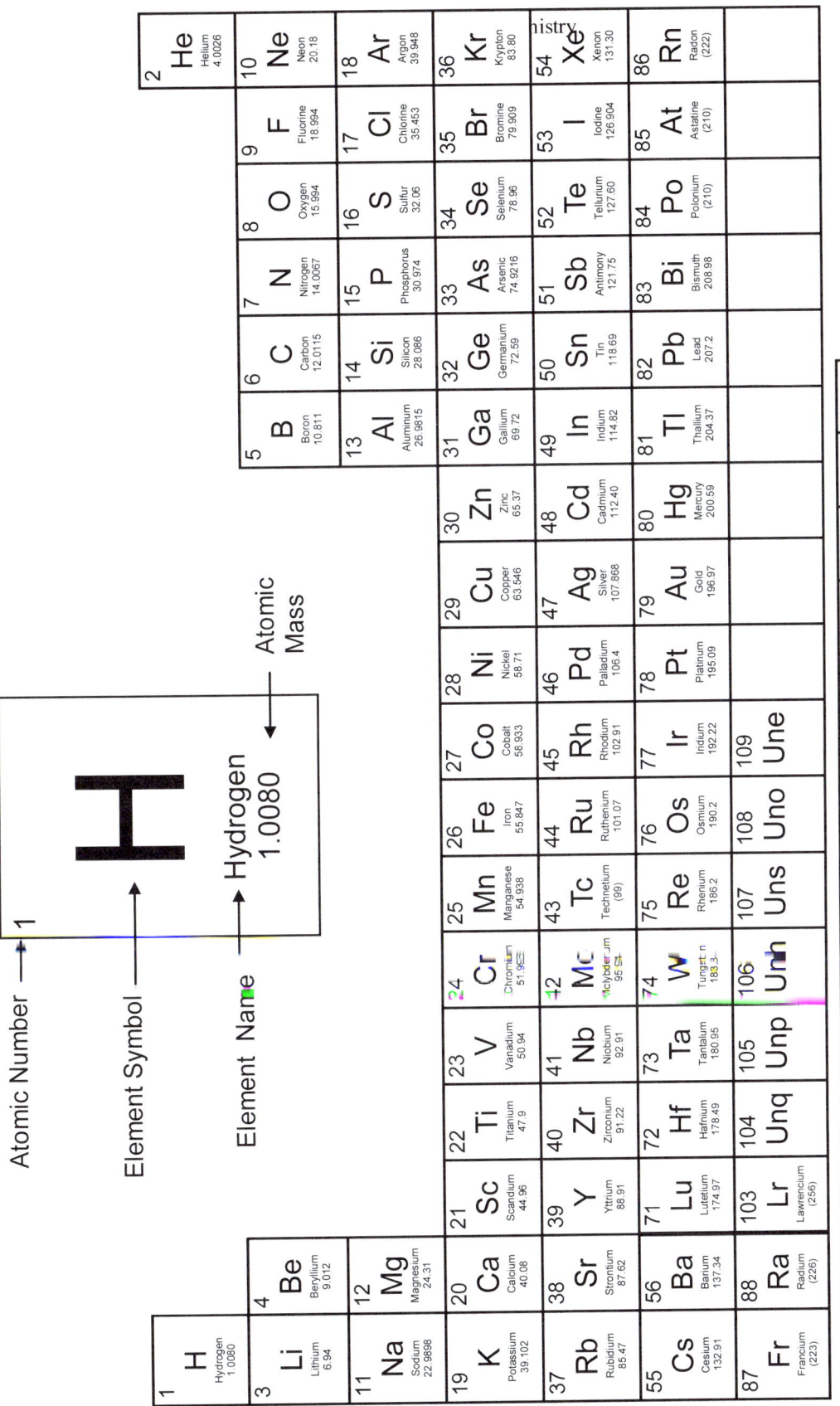

Periodic Table Of the Elements

1 H Hydrogen 1.0080																	2 He Helium 4.0026
3 Li Lithium 6.94	4 Be Beryllium 9.012											5 B Boron 10.811	6 C Carbon 12.0115	7 N Nitrogen 14.0067	8 O Oxygen 15.994	9 F Fluorine 18.994	10 Ne Neon 20.18
11 Na Sodium 22.9898	12 Mg Magnesium 24.31											13 Al Aluminum 26.9815	14 Si Silicon 28.086	15 P Phosphorus 30.974	16 S Sulfur 32.06	17 Cl Chlorine 35.453	18 Ar Argon 39.948
19 K Potassium 39.102	20 Ca Calcium 40.08	21 Sc Scandium 44.96	22 Ti Titanium 47.9	23 V Vanadium 50.94	24 Cr Chromium 51.99	25 Mn Manganese 54.938	26 Fe Iron 55.847	27 Co Cobalt 58.933	28 Ni Nickel 58.71	29 Cu Copper 63.546	30 Zn Zinc 65.37	31 Ga Gallium 69.72	32 Ge Germanium 72.59	33 As Arsenic 74.9216	34 Se Selenium 78.96	35 Br Bromine 79.909	36 Kr Krypton 83.80
37 Rb Rubidium 85.47	38 Sr Strontium 87.62	39 Y Yttrium 88.91	40 Zr Zirconium 91.22	41 Nb Niobium 92.91	42 Mo Molybdenum 95.94	43 Tc Technetium (99)	44 Ru Ruthenium 101.07	45 Rh Rhodium 102.91	46 Pd Palladium 106.4	47 Ag Silver 107.868	48 Cd Cadmium 112.40	49 In Indium 114.82	50 Sn Tin 118.69	51 Sb Antimony 121.75	52 Te Tellurium 127.60	53 I Iodine 126.904	54 Xe Xenon 131.30
55 Cs Cesium 132.91	56 Ba Barium 137.34	71 Lu Lutetium 174.97	72 Hf Hafnium 178.49	73 Ta Tantalum 180.95	74 W Tungsten 183.3	75 Re Rhenium 186.2	76 Os Osmium 190.2	77 Ir Iridium 192.22	78 Pt Platinum 195.09	79 Au Gold 196.97	80 Hg Mercury 200.59	81 Tl Thallium 204.37	82 Pb Lead 207.2	83 Bi Bismuth 208.98	84 Po Polonium (210)	85 At Astatine (210)	86 Rn Radon (222)
87 Fr Francium (223)	88 Ra Radium (226)	103 Lr Lawrencium (256)	104 Unq	105 Unp	106 Unh	107 Uns	108 Uno	109 Une									

57 La Lanthanum 138.91	58 Ce Cerium 140.12	59 Pr Praseodymium 140.91	60 Nd Neodymium 144.24	61 Pm Promethium (147)	62 Sm Samarium 150.4	63 Eu Europium 151.96	64 Gd Gadolinium 157.25	65 Tb Terbium 158.9	66 Dy Dysprosium 162.50	67 Ho Holmium 164.93	68 Er Erbium 167.26	69 Tm Thulium 168.93	70 Yb Ytterbium 173.04
89 Lu Lutetium 174.97	90 Th Thorium 232.0	91 Pa Protactinium 231.0	92 U Uranium 238.03	93 Np Neptunium 237.0	94 Pu Plutonium (244)	95 Am Americium (243)	96 Cm Curium (247)	97 Bk Berkelium (247)	98 Cf Californium (251)	99 Es Einsteinium (254)	100 Fm Fermium (257)	101 Md Mendelevium (258)	102 No Nobelium (255)

Atomic Number → 1
Element Symbol → H
Element Name → Hydrogen
Atomic Mass → 1.0080

Made in the USA
Columbia, SC
28 October 2023